43 Structure and Bonding

Editors:
J. B. Goodenough, Oxford · P. Hemmerich, Konstanz
J. A. Ibers, Evanston · C. K. Jørgensen, Genève
J. B. Neilands, Berkeley · D. Reinen, Marburg
R. J. P. Williams, Oxford

Bonding Problems

With Contributions by
R. Englman   J. C. Green   C. K. Jørgensen
I. Moura   J. J. G. Moura   K. N. Raymond
W. L. Smith and A. V. Xavier

With 58 Figures and 30 Tables

Springer-Verlag
Berlin Heidelberg GmbH 1981

ISBN 978-3-662-15790-9 ISBN 978-3-540-38465-6 (eBook)
DOI 10.1007/978-3-540-38465-6

Library of Congress Catalog Card Number 67-11280

Table of Contents

The Conditions for Total Symmetry Stabilizing Molecules, Atoms, Nuclei and Hadrons

Christian K. Jørgensen

Département de Chimie minérale, analytique et appliquée, Université de Genève,
CH-1211 Geneva 4

Though the totally symmetric (and non-degenerate) wave-function both represents the neutral element of Hund vector-coupling and the origin of the repetitive pattern in the Periodic Table, it has not always low energy, and groundstates having J up to 8 are known. Both in compounds and monatomic entities, Russell-Saunders coupling is usually a good approximation, and classifies the levels correctly (like electron configurations). In nuclei with even Z and N, the groundstate (with I zero) is nearly always followed by an excited state having I = 2 at energies being a function of Z and N clearly connected with M. G. Mayer's shell model. The question of constituents (also α-particles) is further studied in baryons and mesons constructed from quarks. It is still undecided whether quarks and leptons (falling in at least 3 generations) conceivably involve subquarks (such as rishons or preons) and whether an indefinite stratification of further divisibility is suggested by the observed properties. Protons and neutrons are not considered elementary particles any longer.

Table of Contents

1 Electron Configurations and Russell-Saunders Coupling in Monatomic Entities and in Compounds

Monatomic entities M^{+z} consisting of one nucleus (carrying Z times the electric charge e of a proton) surrounded by $K = (Z - z)$ electrons have been one of the major subjects for quantum-mechanical treatment. If the nucleus is treated as a geometrical point, and no attention is paid to its electric multipole moments, nor to its magnetic moments, the energy levels can be characterized by *even* or *odd parity* and by a quantum number J of total angular momentum. If the coordinates $(-x, -y, -z)$ replace (x, y, z) in the total wave-function Ψ (assuming the nucleus at origo), the even Ψ are not changed at all, and the odd Ψ are multiplied by (-1) in each point. For an odd number K of electrons (and also for one electron) J is half an odd positive integer $(1/2, 3/2, 5/2, ...)$ whereas for even K, the J is a non-negative integer.

In the non-relativistic asymptotic limit, where the reciprocal value $(1/c)$ of the velocity of light vanishes, two additional quantum-numbers become valid, the spin angular momentum S and the orbital angular momentum L. Like J, for even K, the former quantum number S is a non-negative integer, whereas S is a half a positive odd integer for odd K. However, in both cases, a higher limit for S is $(K/2)$. On the other hand, L is a non-negative integer for both even and odd K, and there is no higher limit to L. For historical reasons, the various L values have trivial names (which we do not italicize):

$$
\begin{array}{cccccccccccccc}
L = 0 & 1 & 2 & 3 & 4 & 5 & 6 & 7 & 8 & 9 & 10 & 11 & 12 & ... \\
& S & P & D & F & G & H & I & K & L & M & N & O & Q & ...
\end{array}
\tag{1}
$$

and the quantity $(2S + 1)$ is called the *multiplicity* and is pronounced:

even K	odd K	
S = 0 (singlet)	S = 1/2 (doublet)	
1 (triplet)	3/2 (quartet)	
2 (quintet)	5/2 (sextet)	(2)
3 (septet)	7/2 (octet)	
4 (nonet)	9/2 (decet)	

In many cases, when the relativistic effects are not very pronounced, it is possible to identify *terms*, manifolds of adjacent J-levels characterized by a definite combination of S and L. The J-values belonging to a given term can be found by the operation $S \otimes L$ where the *Hund vector coupling* [1] of the two quantum numbers Q_1 and Q_2 has the results

$$
\begin{aligned}
Q_1 \otimes Q_2 &= (Q_1 + Q_2) \text{ or } (Q_1 + Q_2 - 1) \text{ or } (Q_1 + Q_2 - 2) \text{ or } ... \\
&... \text{ or } (|Q_1 - Q_2| + 1) \text{ or } |Q_1 - Q_2|
\end{aligned}
\tag{3}
$$

It is seen that if S is not larger than L, there occur $(2S + 1)$ different J values, stretching from $(L - S)$ to $(L + S)$. It is a general rule that $(2J + 1)$ independent (mutually orthogonal) Ψ represent the same eigen-value with a definite J. It can be seen from recursion formulae that the number of such *states* is $(2S + 1)(2L + 1)$ providing a second reason for calling $(2S + 1)$ the multiplicity. The situation of recognizable terms is called *Russell-Saunders coupling* and the terms are written with the L-symbol from Eq. (1) having $(2S + 1)$ as left-hand superscript. Thus, ^3H combine $S = 1$ with $L = 5$ and is pronounced "triplet-H". In such symbols, J may be added as a right-hand subscript, the three alternatives in the example being $^3H_4$, $^3H_5$ and $^3H_6$. As described by Condon and Shortley[2] the weak relativistic effects can be simplified into "spin-orbit coupling" with definite multiples of the *Landé parameter* ζ_{nl} (for positive l) characterizing each nl-shell (discussed below). In the frequent case of q electrons in *one partly filled shell*, Hund[1] found that the lowest term has the highest possible S_{max} $(= (q/2)$ for the first half of the shell, with q at most $(2l + 1)$, and $= (4l + 2 - q)/2$ for higher q) combined with the highest L compatible with S_{max}. The first-order width of the distribution of $(2S + 1)$ differing J-levels in a term with S_{max} (and L not smaller than S_{max}) is $(L + \frac{1}{2})\zeta_{nl}$ (with opposite sign in the *inverted* terms for q higher than $2l + 1$). According to Hund[1] the groundstate has $J = (L - S_{max})$ for q between 1 and 21. The lowest term of the half-filled shell has $S = J = (2l + 1)/2$ and $L = 0$ corresponding to only one J-level. For q between $(2l + 2)$ and $(4l + 1)$, the groundstate has $J = (L + S_{max})$.

For a monatomic entity containing only one electron, Russell-Saunders coupling is insured by a special situation. Using lower-case letters for one electron, l is like one of the L-values in Eq. (1), and $j = \frac{1}{2}$ when l vanishes, whereas the relativistic effects otherwise separate $j = (l - \frac{1}{2})$ at lower energy from $j = (l + \frac{1}{2})$ at higher. For one electron, the parity is even, when l is an even integer, and odd, when l is odd. It is noted that a given combination of parity and j corresponds to only one l-value.

It is not generally said in text-books that the approximation of *electron configurations* (assigning from zero to $(4l + 2)$ electrons to each *nl-shell*) has many of the same aspects as the approximation of Russell-Saunders coupling. In both cases, the energy levels can be correctly *classified*, and the number of times a given J-value is represented, does not depend on a moderate extent of mixing of electron configurations. In particular, a definite interval of discrete energy levels containing two or three complete configurations contains all the J-levels predicted. The latter situation was originally supposed by the atomic spectroscopists to prevail generally. Thus, it is important for chemists that in neutral atoms of the transition elements, configurations such as $[18] 3 d^q 4 s^2$, $[18] 3 d^{q+1} 4 s$ and $[18] 3 d^{q+2}$ may overlap, and it may be difficult to tell to which configuration a given level belongs. In the example, the three configurations have the same parity (which is a necessary condition for intermixing due to non-diagonal elements of interelectronic repulsion). It was later[3,4] realized that the largest effects of configuration intermixing (excepting special cases such as $1 s^2 2 s^2$ with $1 s^2 2 p^2$ in the groundstate of $K = 4$) are due to the substitution of two nl-electrons with two electrons in an orbital belonging to the continuum (by having positive one-electron energy). If the l value is different, the radial extension of this orbital is roughly same as the nl-orbital. If the two l values are identical, the continuum orbital has a radial node in the middle of the nl-radial function (in order to remain orthogonal). This produces far larger energetic effects than e.g. the mixing

3

of [18] 3 d^q with [18] 3 d^{q-2} 4 d^2. We write closed-shell K-values in rectangular parentheses. It is frequent to write [Ar], [Kr], [Xe], ... for [18], [36], [54], ... but since many closed-shell systems with positive charge, such as [28], [68], [78], ... are not represented by groundstates of neutral atoms, it may be more convenient always to use the K-numbers. The groundstate of gaseous M^{+2} (with the five exceptions M = La, Gd, Lu, Ac and Th), all M^{+3}, M^{+4}, M^{+5} and M^{+6} (but [5] not certain z values above 6) belongs to the configuration obtained by the *Aufbauprinzip* [1] of filling the shells in the consecutive order

$$1\,s \ll 2\,s < 2\,p \ll 3\,s < 3\,p \ll 3\,d < 4\,s < 4\,p \ll 4\,d < 5\,s < 5\,p \ll$$
$$\ll 4\,f < 5\,d < 6\,s < 6\,p \ll 5\,f < 6\,d\,... \tag{4}$$

where the double inequality signs indicate the closed-shell systems isoelectronic with the noble gases. In text-books are frequently mentioned a similar series for the groundstate of neutral atoms, where 4 s, 5 s, 6 s and 7 s have moved down to follow immediately after the double inequality signs corresponding to K = 18, 36, 54 and 86. However, such an Aufbauprinzip for neutral atoms is of much less interest than Eq. (4), since twenty exceptions occur among the 99 atoms from hydrogen to einsteinium.

It is an experimental fact [6] that the lowest 20 to 400 J-levels of a given mon-atomic entity can be classified by electron configurations (usually either the same as the groundstate, or obtained by letting one or two electrons change their nl-shell) and there is no clear-cut case of any low-lying level being supernumerary to the low-lying configurations expected. When Russell-Saunders coupling is a good approxima-tion, a single partly filled shell can at most have S = $(1 + \frac{1}{2})$ (when half-filled) such as S = $\frac{5}{2}$ for d^5 and $\frac{7}{2}$ for f^7. A few terms are known with S = 5, e.g. belonging to [54] 4 f^7 5 d 6 s 6 p of the gaseous gadolinium atom [7]. However, because of the great stability of closed inner shells, no discrete levels of monatomic entities containing more than two electrons have S as high as (K/2).

It should not be neglected that even ^{1}S terms are not restricted to closed-shell K-values. If *one* partly filled shell contains two or 4 l electrons, (2 l + 1) terms occur, among which ^{1}S has the *highest* energy. If the partly filled shell contains 4, 6, 8, ..., (4 l − 2) electrons, two or more terms ^{1}S occur, most frequently with rather high energy compared with the other, numerous terms. This is not a question of L = 0 since f^3 and f^{11} have the two first excited terms (at the same energy in Racah's theory) ^{4}S and ^{4}F slightly above the lowest term ^{4}I (agreeing with Hund's rules). Said in other words, ^{1}S belonging to configurations involving partly filled shells are destabilized, compared with closed-shell situations. This does not prevent that con-figurations with two partly filled shells may have energies [4] below closed-shell config-urations, e.g. [18] 3 d4 s in Sc$^+$ below both [18] 3 d^2 and [18] 4 s^2, and [18] 3 d^9 4 s in the gaseous nickel atom below [18] 3 d^{10} and the average energy of [18] 3 d^8 4 s^2 (to which the groundstate belongs).

In molecules and polyatomic complex ions, the equilibrium positions of the nuclei determine point-groups [8] which are finite (one of the seven cubic groups, or belonging to one of the seven series D_{nh}, D_{nd}, D_n, C_{nh}, C_{nv}, C_n and S_{2n} including the isolated plane of symmetry C_s, the isolated centre of inversion C_i and, finally, C_1

having only identity as element of symmetry) except when two or more nuclei all are colinear, exemplifying the linear point-groups $D_{\infty h}$ with, or $C_{\infty v}$ without, a centre of symmetry (inversion). If Russell-Saunders coupling is valid in such polyatomic entities, the total spin quantum number S is combined with Λ in the linear point-groups or the symmetry type Γ_n (replacing L in monatomic entities) in the finite point-groups.

Organic chemists emphasize that the large majority of molecules have a ground-state with vanishing S (hence being diamagnetic, or at the most, temperature-independent paramagnetic). They call species with positive S "free radicals". It is inevitable that molecules containing an odd number of electrons (such as NO, O_2^+, O_2^-, ClO_2 or O_3^-) have positive S (and the examples given are not known to dimerize) but it is more striking that the groundstate of O_2 has S = 1. The paramagnetism of the oxygen molecule (which was discovered by Faraday) was rationalized by Lennard-Jones on the basis of *molecular orbital* (M.O) theory, the two loosest bound M.O. (able to accommodate four electrons, as they do in the diamagnetic peroxide anion O_2^{-2}) have exactly the same energy for group-theoretical reasons, much like the (2l + 1) orbitals of a nl-shell in spherical symmetry. In polyatomic entities, identical (or almost identical) M.O. energies have many of the same consequences as Hund's rules for a monatomic entity with an electron configuration containing a partly filled shell.

At this point, there is an interesting graduation from p over d to f group compounds[8,9]. Even when p-like orbitals can be recognized, the systems with an even number of electrons generally have S = 0. Typical cases are the (K = 32) bromine(III) and (K = 50) iodine(III) and xenon(IV) complexes BrF_4^-, ICl_4^- and XeF_4 containing one lone-pair perpendicular on the molecular plane. In d-group complexes, both high-spin (S according to Hund's rule) and low-spin behaviour is known. For instance, d5 systems can have S = $\frac{5}{2}$ or $\frac{1}{2}$ in their groundstate, and d6 systems S = 2 or 0. These examples are all four compatible with octahedral symmetry (compare hexaqua and hexacyano complexes of iron(II) and iron(III)). Other cases, such as (K = 26) 3 d8 nickel(II) complexes have a marked correlation with the stereochemistry. Thus, the groundstate of octahedral chromophores, such as $Ni(II)O_6$ in $Ni(OH_2)_6^{+2}$, $Ni_xMg_{1-x}O$, undiluted NiO and $Ni(II)N_6$ in $Ni(NH_3)_6^{+2}$ have S = 1 because the two anti-bonding 3 d-like orbitals in "ligand field" theory contain 2 (and not 4) electrons (and the three non-bonding 3 d-like orbitals are filled) whereas diamagnetic behaviour of nickel(II) is found in quadratic, rectangular or tetragonal-pyramidal chromophores (such as $Ni(CN)_4^{-2}$, $Ni(S_2P(OC_2H_5)_2)_2$ or $Ni(CN)_5^{-3}$, respectively) with one of the five 3 d-like orbitals empty. The "ligand field" description had the (somewhat unexpected) consequence that the *isoelectronic series*[5] introduced by Kossel in 1916 where the K-values derived from Eq. (4) for monatomic species

$$2 \leqslant 4 < 10 \leqslant 12 < 18 \leqslant 28 < 30 < 36 \leqslant 46 < 48 < 54 \leqslant$$
$$\leqslant 68 < 78 < 80 < 86 \leqslant 100 < \dots \tag{5}$$

were found (with exceptions of K = 4, 12 and 30) to correspond to six to thirteen consecutive oxidation states, such as K = 10 from C(–IV) to Cl(VII) or K = 46 from Ru(–II) to Xe(VIII), whereas other K-values rarely are represented by as many as four or five consecutive oxidation states. Actually, all the K-values from 19 to 100 are known in at least one well-defined oxidation state of a mono-meric complex (excluding catenation between identical atoms, such as F_3CCF_3, O_2^{-2}, $O_3SSO_3^{-2}$, F_5SSF_5, $(OC)_5MnMn(CO)_5$, $(NC)_3NiNi(CN)_3^{-4}$ and Hg_2^{+2}) distinct-ly different from the definition [4] of oxidation numbers. Whereas all known cases of K = 21 show groundstates (with $S = \frac{3}{2}$) belonging to a M.O. configuration with three (essentially non-bonding) d-like electrons, we already mentioned that K = 23 with five d-like electrons may show either $S = \frac{5}{2}$ (like gaseous Mn^{+2} and Fe^{+3}) or $S = \frac{1}{2}$. The characteristic difference between the chemistry of elements outside the transi-tion groups at one hand, and the d and f groups on the other hand, is that K (when it is defined) nearly always is an even integer in the former case, whereas the most stable (or most frequent) oxidation state of a transition element readily can have an $\cdot$ odd K-value, such as Cr(III), Mn(II), Fe(III) or all the trivalent lanthanides from Ce(III) to Yb(III) with even Z.

In the 4 f group, the situation is rather different from the d-groups, since the deviations from spherical symmetry are smaller [10] than the first-order relativistic effect "spin-orbit coupling". Hence, the J-values of the groundstate and of nearly all the excited states (excluding accidental near-degeneracies) of [54] $4f^q$ remain well-defined, like in monatomic entities. This is true, not only for transparent, isolating compounds (such as MF_3 or M_2O_3) and diluted crystals (such as $M_xLa_{1-x}Cl_3$ and $M_xY_{1-x}VO_4$) and glasses, as well as aqueous solutions [11,12], but also in black, low-energy semi-conductors such as MSb and in metallic alloys and elements. In the latter case, the *conditional oxidation state* [4] M[III] is said to occur, when magnetic or other physical properties indicate the number q = (Z − 57) of electrons in the 4 f shell, whereas M[II] is defined by q = (Z − 56). The photo-electron spectra [13,14] of antimonides and the metallic elements indicate the J-level structure of the M[IV] $4f^{q-1}$ obtained by X-ray ionization of M[III]. The typical separation of the seven one-electron energies of 4 f group compounds [10] is 400 to 800 cm^{-1} (0.05 to 0.1 eV) which is also the order of magnitude of the width of the distribution of (2 J + 1) states in each level. Finer details of the magnetic behaviour [15] of metallic alloys and elements show that the (2 J + 1) states of the lowest M[III] level are separated about half as much as in anhydrous lanthanum chloride.

For reasons becoming apparent in the next section, it is interesting to note that lanthanides in condensed matter are sufficiently close to spherical symmetry that Hund's rules all apply to the partly filled 4 f shell. Thus, the highest J-value for the groundstate is 8 (combined with S = 2 and L = 6) in $4f^{10}$ holmium(III) and dyspro-sium(II). Gaseous Ho^{+3} and Dy^{+2} also have this groundstate $^5I_8$ (consisting of 17 states) belonging to [54] $4f^{10}$. It may be noted that the gaseous dysprosium atom [7] also has this groundstate, but in this case belonging to the configuration [54] $4f^{10}$ $6s^2$.

In the transthorium elements (with Z above 90) there was around 1955 a con-troversy [16,17] whether some 6 d electrons might occur, rather than a partly filled 5 f shell in agreement with Eqs. (4) and (5). The situation seems quite complicated in alloys of protactinium, uranium and neptunium with other metals, whereas pluto-nium may choose between 5 f^4 Pu[IV] and 5 f^5 Pu[III] in the individual alloys or modifications of the element. On the other hand, americium and curium seem always to occur as $5 f^6$ Am[III] and $5 f^7$ Cm[III] in the metallic state. It is well-established[16,18] that isolating compounds of transthorium elements contain $(Z - 86 - z)$ electrons in a partly filled 5 f shell when the oxidation state is + z. The "ligand field" separations between the one-electron energies of the seven f-like orbitals are known[19] to be 0.5 to 1 eV in hexahalide complexes of $5 f^1$ Pa(IV), $5 f^2$ U(IV) and $5 f^1$ U(V), approaching the values (1 to 3 eV) typical for the 3 d group. On the other hand, $5 f^3$ U(III) recently studied[20] in $U_x La_{1-x} Cl_3$ have "ligand field" effects only twice as large as the homologous $4 f^3$ Nd(III), and similar regularities are observed in the subsequent 5 f group M(III). It is very difficult to oxidize $5 f^{14}$ nobelium(II) representing the closed-shell K = 100 in Eq.(5). It is likely that lawrencium (Z = 103) and immediately following elements most frequently have the oxidation state $(Z - 100)$ but they soon begin to form certain complexes with a partly filled 6 d shell. The predicted chemistry of heavier elements (up to Z = 184) has been discussed at length[16] and the inner shells are strongly influenced by relativistic effects[21,22]. However, the chemistry is much less modified from the non-relativistic behaviour, and mostly by the strong stabilization of 7 s and 8 s orbitals and of the third of the 7 p and 8 p shells having $j = \frac{1}{2}$.

Professor Robert Englman was so kind as to point out to the writer that "non-degenerate groundstate" might be a more suitable word than "totally symmetric groundstate". The argument is that a set of degenerate states belonging to the same eigen-value of the Schrödinger equation behaves collectively as having total symmetry. It is very difficult to say a clear-cut "yes" or "no" to such a suggestion. A simplified analysis can be made of the question whether the boron atom in its lowest configuration $1 s^2 2 s^2 2 p$ has spherical symmetry or not. There is no doubt that the six degenerate states form a spherically symmetric basis set, but at the same time, they form three Kramers doublets constituting the zero-order wave-functions for a uniaxial perturbation, such as the linear "ligand field" in the chromophore XBX with long B-X internuclear distances[8]. Hence, an arbitrarily weak non-spherical (and non-cubic) perturbation is capable of producing energy differences between certain of the 6 states. The situation is intrinsically different in the ^1S groundstate of the beryllium atom, independently of the numerical question of the squared amplitude of the next-largest contribution $1 s^2 2 p^2$ to the total wave-function written as a mixture of electron configurations, where $1 s^2 2 s^2$ is the predominant contribution. "Accidental" degeneracy between non-totally symmetric wave-functions can have rather unexpected consequences; Epstein was the first to point out that the coinciding energies of 2 s and 2 p (of opposite parity) in the non-relativistic treatment of the hydrogen atom (and other monatomic entities with one electron) produce an electric dipole moment, which is otherwise impossible for a system having a centre of inversion. As far we know[11], the smallest distance to a low-lying J-level with opposite parity of the groundstate in a neutral atom occurs in the lowest level of [54] $4 f^8 5 d6 s^2$ at 286 cm^{-1} above the groundstate of the gaseous terbium atom belonging to [54] $4 f^9 6 s^2$. In Th^{+2}, the lowest level of [86] $6 d^2$ occurs only 63 cm^{-1} above the groundstate belonging to [86] 5 f 6 d. Such closely adjacent levels of opposite parity would show an apparent electric dipole moment saturating at high electric field strengths. Englman[23] argues that the Jahn-Teller effect in polyatomic systems having degenerate sets of groundstates in a given high symmetry may be predetermined by properties of the individual atom, such as boron compared with beryllium. However, it is an empirical fact of great importance for the chemistry of post-transitional elements[8,24] that many molecules and polyatomic ions refuse to exhibit the highest symmetry available to them.

In spherical symmetry (of monatomic entities) there is only one type of non-degenerate level not possessing total symmetry: J zero with odd parity. This is, for instance, true for the first excited state of the mercury atom (and the isoelectronic gaseous ions Tl^+, Pb^{+2} and Bi^{+3}) belonging to [78] 6 s 6 p which plays an interesting rôle in the luminescence[10,25] of bismuth(III) in condensed matter. In the limit of Russell-Saunders coupling, the (rather unfrequent) non-degenerate term ^1S combined with odd parity is predicted in the excited configuration 1 s² 2 s 2 p² 3 p of the carbon atom. A much less hypothetical question is whether a chromophore in a finite point-group may show a non-degenerate groundstate, which is not totally symmetric. Many point-groups[8] certainly allow such states ($^1A_2$, $^1B_1$, $^1B_2$, ... in contrast to $^1A_1$). A naïve M.O. interpretation suggests that such states cannot be groundstates. If x (absolutely or almost) degenerate M.O. contain less than 2 x electrons, Hund's rules seem universally valid that a state with positive S has lowest energy. But if they contain 2 x electrons, the totally symmetric $^1A_1$ is the only state obtained. This does not prevent[7] that the ground level (with J = 4) of the gaseous cerium atom (belonging to [54] 4 f 5 d 6 s²) and the lowest odd level of Ce^{+2} (belonging to [54] 4 f 5 d) has more $^1G_4$ than $^3H_4$ character. It is not known whether this exception from Hund's rules for *two* partly filled shells in a monatomic entity may have an analogy in M.O. configurations. It is interesting to note[8] that 3 d group complexes showing *pronounced* Jahn-Teller distortions[26] have unbalanced occupation of M.O. in the high symmetry, by having 0 and 1, or alternatively 1 and 2, electrons in two M.O. which would have the same energy, e.g. in regular octahedral MX_6. However, tetragonally elongated 3 d⁴ chromium(II) and manganese(III) complexes have $^5A_1$ and 3 d⁹ copper(II) $^2B_1$ groundstates, and distinctly not $^5B_1$ nor $^2A_1$.

2 Nuclear Structure and Properties

Molecular spectroscopists and chemists studying polyatomic systems tend to consider nuclei as geometrical points. In this model, the instantaneous picture[8] of a homonuclear diatomic molecule is invariantly of the symmetry $D_{\infty h}$ whereas a heteronuclear diatomic system cannot loose the symmetry $C_{\infty v}$ and a system with three nuclei cannot have a lower symmetry than the point-group C_s having a plane of symmetry as only element of symmetry besides identity. In the strong sense in which the overwhelming majority of all real numbers are irrational (as shown by Cantor, the cardinality of the rational numbers, and even of the algebraïc roots of polynomials with rational coefficients, is not higher than the denumerable set of integers) an instantaneous picture of four or more geometrical points *almost* always has the lowest possible symmetry C_1. The vibrations of the nuclei (accepting three degrees of freedom of translation to be disregarded, assuming Born-Oppenheimer separability of the wave-function, as well as three degrees of rotational motion in the case of three or more nuclei, and two degrees in diatomic molecules) correspond to a distribution of instantaneous pictures scattered around a set of equilibrium positions, frequently representing a point-group of high symmetry. The cubic point-group T_d found in CH_4, SiF_4, $CoCl_4^{-2}$, ... is only such a mean value of the nuclear positions (this restriction is accentuated in X-ray diffraction of crystals, where only the time-average picture of the average content of the unit cell is determined). Like in molecular absorption spectra, the Franck-Condon principle is satisfied by photo-electron spectra, where the "vertical" ionization energies refer to the average internuclear distances, though a few molecules such as methane [27] show some additional structure of the first photo-electron signal (corresponding to removal of one of the six electrons in the three degenerate M.O. consisting of C2p and H1s in the L.C.A.O. model) is due to CH_4^+ and CD_4^+ being Jahn-Teller unstable. For our purpose, it is important to note that the most precisely known [28] internuclear distances R with an uncertainty of the order of magnitude 10^{-5} are obtained from rotational spectra (in the micro-wave region) with energy levels $J(J + 1) \langle R^{-2} \rangle$ times an expression dependent on the atomic masses and on constants of Nature. In diamagnetic diatomic molecules, the quantum number J is a non-negative integer. The question of time-scale once more shows up in this evaluation, the average value of R^{-2} of the instantaneous pictures being the precisely determined quantity. Actually, the typical scattering of the instantaneous R is several percent around the average value. In polyatomic molecules, three moments of inertia can be derived from rotational spectra, corresponding to the principal axes of a general ellipsoid. If isotopic substitution can be performed on a polyatomic molecule, the slightly different sets of each three moments of inertia can be used, in fortunate circumstances, to evaluate all the internuclear distances (in the sense defined above). By the way, if the difference between optically active enantiomers is neglected, the manifold of internuclear distances suffice to define the nuclear skeleton, without any reference to the explicit point-group.

The reason why nuclei for many purposes can be approximated by geometrical points is their very small diameters, compared with the internuclear distances. Though the nuclear surface [29] is not absolutely sharply defined, the nuclear matter has a roughly constant density close to 10^{14} g/cm^3 and hence, the radius of a nucleus with the atomic mass number A is 1.2 fm (1 fermi = 1 femtometer = 10^{-13} cm) times the cube-root of A. The slightly different volumes of different isotopes of the same element produce shifts of atomic spectral lines (mainly connected with the number of s electrons in the electron configurations of the excited and of the lower level) of the order of magnitude 10^{-5} times the wave-numbers. However, a much more conspicuous isotope effect on spectra of monatomic entities (a comparable structure can be resolved in diatomic molecular spectra) occurs when the nuclear groundstate has a positive spin quantum number I. This quantity is zero for all stable nuclei, if both A and the atomic number Z are even integers. The *hyperfine structure* of atomic spectra can spread over several cm^{-1}, i.e. around 10^{-4} times the wave-numbers, and corresponds to a new quantum number $F = I \otimes J$ obtained by vector-coupling defined in Eq.(3). When J is at least as large as I, a set of $(2I + 1)$ adjacent energy levels is observed. A more direct technique of observing such levels is *nuclear magnetic resonance* (studying the Zeeman effect due to an external magnetic field) and also the hyperfine structure (introduced by nuclei having positive I) of electron magnetic resonance of paramagnetic species [30] providing a rather objective technique of evaluating L.C.A.O. delocalization coefficients. Another property of certain nuclei of interest for chemists is the electric quadrupole moment (only having observable consequences, if I is at least 1) allowing *nuclear quadrupole resonance* [31] to be observed between the (closely adjacent) energy levels obtained in an external uni-axial perturbation from neighbour atoms.

Besides the experimental fact that some nuclei (those with positive I) have properties (such as magnetic moments) somewhat incompatible with being geometri-cal points, the evidence obtained from radioactive isotopes strongly suggested some kind of composition of the nuclei. Since the time of Crookes and Thomson, electrons were recognized to be a constituent of all matter. They had been shown to posses the very low atomic weight 0.00054858 (1 unit of atomic weight has $m_0 c^2 =$ 931.50 MeV) when moving at low velocities. It is easy in a laboratory to let electrons through a potential difference 511 000 V after which their velocity is $(\sqrt{3}/2) c$ and their inertial mass twice as large as their rest-mass m_0. The old hypothesis of Prout of hydrogen being the primordial element was reformulated as protons (a hydrogen atom, including its electron, has the atomic weight 1.007825 relative to an atom of the abundant carbon isotope equal to 12) being the only other, and "massive" constituent, the nuclei "consisting" of A protons and $(A - Z)$ electrons. This model prevailing before 1930 had a great impact on chemists; the two major types of chemical reactions are "redox" (reduction-oxidation) transferring electrons (like phlogiston) and Brønsted acidity, transferring protons. However, this model ran into a difficulty even before the discovery of the neutron. The nitrogen 14 nucleus has a groundstate with I = 1, but it should contain 14 protons and 7 electrons, altogether 21 *fermions* which should produce a system with quantum numbers (such as I or J) being odd positive integers divided by 2. Quite generally, quantum mechanics requires a set of bosons, to combine to a system (with I or J being a non-negative integer) also obeying Bose-Einstein statistics, whereas an *odd* number of fermions (with intrinsic half-numbered spin) obeys Fermi-Dirac statistics, but an *even* number of fermions represent Bose-Einstein statistics. Text-books usually introduce a third "elementary" particle, the *neutron*, and the nucleus is said to contain Z protons and $N = (A - Z)$ neutrons. Thus, nitrogen 14 nuclei are bosons because A is even. The observable properties of the free neutron are slightly disturbing for such a simple picture; it has a magnetic moment about $(-\frac{2}{3})$ of that of the proton, though it has no electric dipole moment ($I = \frac{1}{2}$ prevents higher multipole moments from being observ-able). The *anti-particles* were established a few years after the neutron was detected. All fermions and some bosons (exceptions are the photon and the neutral pion) have anti-particles with opposite electric charges and magnetic moments. The free neutron is radioactive (with the half-like 10.6 min and the 1.4427 times longer average life-time 15.3 min) forming a proton, an electron and an anti-neutrino (to be discussed below) whereas the anti-neutron is distinctly different by decaying (with the same half-life) to an anti-proton, a positron and a neutrino. The comparatively slow decay of the neutron was not felt to disqualify it as an elementary particle; once, the neutron is incorporated in a stable nucleus, it is no longer radioactive, and it became usual to say that the nucleus consists of A *nucleons* among which Z are in the state of protons, and N = (A − Z) in the state of being neutrons. However, once the doors were opened for radioactive "elementary" particles, they proliferated, and became nearly as numerous as the hundred elements in the Periodic Table. We return to these entities after having discussed problems more directly related to the structure of nuclei.

Some of the radioactive isotopes isolated from thorium and uranium minerals perform exponential β-decay, i.e. they emit electrons, of which the kinetic energy can be in considerable excess of $m_0 c^2$. Such β-rays have been of great help in verifying the predictions of the special theory of relativity, and provided an obvious argument in favour of (A − Z) electrons occuring inside the nucleus. However, subsequent experience clearly demonstrated that the emission of a particle is no proof of its *pre-existence* in the nucleus. The two major problems with β-decay is that the kinetic energies of the electrons are spread over a large interval, from very small values to a limiting maximum (which could be identified with the energy difference between the groundstates of the original and of the product isotopes corrected for $m_0 c^2$ of the electron) and another being that emission of *one* fermion is not compatible with A remaining even (or odd) during the β-decay, without any change-over between Bose-Einstein and Fermi-Dirac statistics. A third problem is that the kinetic energy of a particle as light as an electron confined in a volume as small as a nucleus would be exceedingly high.

Pauli suggested in December 1930 to remove these three problems by the simultaneous emission of an electron and (what we now call) an anti-neutrino. In 1934, Fermi gave a quantitative theory for the probability of sharing of kinetic energy between the two particles. Cases (such as tritium decaying to helium 3) of marginally low energy of decay available indicate *very* much smaller rest-mass of the anti-neutrino than that of the electron. It is generally agreed today that the behaviour of neutrinos and anti-neutrinos is much more comprehensible, if their rest-masses are exactly zero (like the photons) having the corollary that they are seen by all observers to move with the velocity c. The first artificial radioactive isotope (characterized by Irene and Frederic Joliot-Curie in 1934) was phosphorus 30 (half-life 2.5 min) forming the stable silicon 30 by emitting a positron (soon annihilating together with an electron in the surrounding matter) and a neutrino. An alternative to positron emission is *electron capture* where an electron (normally present in the 1 s shell closest to the nucleus) is absorbed by the radioactive nucleus, at the same time as a neutrino is emitted. It is amusing for ecologically minded persons that a-third of the radioactivity in the Earth's crust (and nearly all their own) is due to the isotope K 40 present with an abundance 0.012 percent in potassium. It is energetically unstable both with respect to argon 40 (explaining why one percent of the atmosphere consists of this isotope, far more abundant than both the other four noble gases, and argon 36 and 38) and calcium 40. With a half-life of 1277 million years (corresponding to 16 times higher abundance 5 milliard years ago), it emits (with a probability 89.3 percent) electrons (and anti-neutrinos) to form the most abundant calcium isotope 40, it undergoes electron capture (with the probability 10.7 percent) to form argon 40, whereas the probability of emitting a positron (and a neutrino) is only 0.001 percent [32].

The most striking radioactive decay among the isotopes found in thorium and uranium minerals is the emission of α-*particles*, i.e. helium 4 nuclei. If the arguments about pre-existing entities in nuclei were valid, this would suggest that nuclei are systems containing α-particles. However, with present-day understanding of the quantum mechanics, it is more a question of a nucleus characterized by the proton and neutron numbers (Z, N) dissociating to two or more (spallation) products characterized by $(Z_1, N_1), (Z_2, N_2), ...$ conserving $Z = Z_1 + Z_2 + ...$ and $N = N_1 + N_2 + ...$ The time-dependent Schrödinger equation describes the rate strongly dependent on the height of the activation barrier, and the empirical findings of Nuttal and Geiger that the logarithm of the half-life of α-decay is a rapidly varying, linear function of the penetration range of α-particles in air, were rationalized by Gamov into a similar logarithmic dependence on the kinetic energy of the emitted α-particle [33]. A few short-lived fission products with unusually high N (for their Z) emit neutrons with a half-life of a few seconds (these "delayed neutrons" have an enormous importance for reactor technology) and a few nuclei with unusually low N (such as lithium 5, boron 9 or aluminium 23) rapidly emit protons. However, such an emission of one of the two nucleons is rarely energetically feasible. In the interval of A-values between 2 and 11, helium 4 is by far the most stable nucleus, and since it is so relatively light, spallation producing carbon 12 or heavier products has far longer half-life. This does not prevent that for sufficiently high Z-values, other channels of dissociation become important. The nuclear reaction *fission* of uranium 233 or 235 with slow neutrons (where the binding energy of the neutron to form the isotopes 234 or 236 supplies the activation energy needed) is a division (evolving some 200 MeV) into two nuclei with somewhat differing A. The probability distribution (illustrating the many channels utilized) has a shape of a camel back with maxima close to A = 95 and 138. At the same time, two or three "instantaneous" neutrons are emitted. The first nucleus where *spontaneous fission* has been well-established, is uranium 238 (constituting 99.3 percent of uranium in minerals at present) where one nucleus out of 1835000 α-emitters undergoes fission without precedent neutron capture. Since the half-life for α-decay of this nucleus [32] is 4468 million years, it means that the half-life for spontaneous fission would be $8.2 \cdot 10^{15}$ years, if it was the only mode of decay. In heavier nuclei, spontaneous fission becomes a predominant alternative; thus, the half-life is 2.64 years for californium 252 (however, 96.9 percent is due to α-decay), 60 days for californium 254 (99.7 percent spontaneous fission), 158 min for fermium 256 (92 percent spontaneous fission) and 0.08 s for (Z = 104, A = 260).

Much like two helium atoms do not combine to a diatomic molecule, two α-particles do not form a stable beryllium 8 nucleus (which is known to divide within 10^{-16} s). On the other hand, 3, 4, ..., 10 α-particles oligomerize to the stable nuclei carbon 12, oxygen 16, . . ., calcium 40 which are each time the most abundant isotope of their element (with exception of argon 36 discussed above). Also the abundances [34] in the normal stars (such as our Sun) strongly point to a pronounced stability of these oligomers, though the most frequent isotope, by far, is the free hydrogen 1, and about 8 percent of the atoms (and 25 percent of the total mass) is helium 4, of which a major part was formed a few minutes after the "Big Bang", the singularity from which the Universe evolved some 10^{10} years ago [35] and the rest of the helium is the product of the source of energy (by hydrogen fusion) of ordinary stars. Though the Sun only transmutes $9 \cdot 10^{-12}$ of its mass from hydrogen to helium per year (at the moment), other stars are much more prodigalous and can show transmutation rates well above 10^{-9} year $^{-1}$ indicating a duration much shorter than of the Universe. There are good arguments (including the relative abundances of radioactive isotopes with half-lifes in the 10^9 year class) that both the Sun, the Earth and its Moon have condensed as recently as 5 milliard years ago. It is obvious [34,36] that the relative abundances of elements and of their isotopes are determined by kinetics; thermodynamical equilibria at temperatures [35] below 10^9 degrees would produce elements such as iron and nickel; neutron stars with A above 10^{57} and their implosion products, the black holes. [37] The two latter types of entities are determined by the gravitational attraction, which is negligible inside atoms and nuclei, but proportional to $A^{5/3}$ for a constant density of the neutron fluid.

The A-values 2 and 3 are not represented by any strongly bound nuclei, and A = 5 not at all (hydrogen 5 and helium 5 immediately loosing a neutron, and lithium 5 and subsequent isotopes a proton). The binding energies (relative to protons and neutrons) are in MeV for the stable light nuclei:

deuterium	2.225	beryllium 9	58.165	
helium 3	8.500	boron 10	64.751	
helium 4	28.296	boron 11	75.423	(6)
lithium 6	31.994	carbon 12	92.163	
lithium 7	39.244	oxygen 16	127.621	

They are seen to increase more rapidly than the number of mutual interactions $A(A-1)/2$ for A = 2, 3 and 4 and reaching a pronounced saturation in helium 4. Then, the pattern seem to repeat for the two next multiples of 4, the sum of the binding energy 8.482 MeV of tritium and 28.296 MeV of helium 4 being 2.466 MeV below that of lithium 7. By the same token, a triton and two α-particles have their binding energies 10.349 MeV lower than of boron 11. The trimerization energy of α-particles is seen to be 7.275 MeV and the tetramerization energy to oxygen 16 is 24.437 MeV, again showing a ratio 1 : 3.329 closely similar to the ratio 3.359 between the binding energies of helium 3 and 4. On the other hand, there is distinctly no tendency of oxygen 16 to tetramerize to germanium 64 or the other nuclei with A = 64 formed by electron capture or positron emission from this nucleus.

It is the general consensus among nuclear physicists [38] that the "unsaturated" behaviour of nuclei with A below 12 in Eq. (6) is atypical, in sofar the *liquid drop* model becomes a reasonable approximation in heavier nuclei. It may be noted that the smallest drop consisting of A identical spheres, where at least one particle is not in the surface, occurs for (both cuboctahedral and icosahedral) A = 13, related to the fact that $4\pi = 12.56637...$ This model was proposed by C.F. von Weiszäcker in 1935, and one example [39] of the parametrization of the total atomic weight in the unit of 0.001 chemical unit is

$$1008.6650\,A - 0.8400\,Z - 16.72\,A + 18.5\,A^{2/3} + 100\,(A - 2\,Z)^2/4\,A + $$
$$+ \; 0.75\,Z^2/A^{1/3} + (-1)^{Z+1}\,\delta(\text{even}\,A) \tag{7}$$

The corresponding unit is 0.93150 MeV. Since isotopes occur together in definite elements (separated from minerals by chemical techniques), it is the tradition to write Eq.(7) as a function of A and Z, though an equivalent expression can be written with N and Z as variables. The two first terms express the atomic weights of the "constituents" the N neutrons (1.0086650) and the Z hydrogen atoms (1.0078250), so the five last terms of Eq.(7) represent the predicted binding energy relative to Z protons and N neutrons. The *bulk term* 16.72 A represents the stabilization of A nucleons, disregarding their character as protons or neutrons. The *surface term* 18.5 A$^{2/3}$ represents (in some averaged fashion) the destabilization of nucleons situated in the surface layer of the nucleus, lacking the binding to any nucleons outside the surface. Going from A = 27 to 216, the surface term cancels from 38 to 19 percent of the bulk term. The *asymmetry term* proportional to $(N - Z)^2/A$ indicates the destabilization connected with differing N and Z, and is the driving force behind β-radioactivity. Originally, the asymmetry term was considered to express a second-order effect in the interaction between identical nucleons, but there is little doubt that a large proportion is due to unfavourable use of the proton shells discussed below, and hence in a sense connected with Pauli's exclusion principle. The sixth contribution to Eq. (7) is the *Coulomb term* proportional to Z^2 divided by the nuclear radius. The last contribution is the "even-odd term" or *alternating term* vanishing for odd A. The alternating term represents a stabilization for even Z combined with even N, and a destabilization (of opposite sign) for odd Z combined with odd N. The parameter δ (even A) is close [40] to 13 millimassunits divided by the square-root of A.

The numerical parameters 16.72; 18.5; 100 and 0.75 millimassunits in Eq.(7) can easily be modified to somewhat different sets because the binding energy per nucleon is a very shallow function of A. Thus, the binding energy (divided by A) is 7.68 MeV for carbon 12 and 7.79 MeV for oxygen 16 in Eq.(6), and increases to a maximum at 8.8 MeV around iron 56. This may be interpreted as weakly decreasing importance of the surface term. The slow decrease after this maximum down to 7.47 MeV for californium 250 is essentially due to increasing predominance of the Coulomb term of interprotonic repulsion. It is noted that the binding energy per nucleon remains between 48 and 57 percent of the bulk energy in the whole interval from A = 12 to 250, and is quite exactly half the bulk energy in oxygen 16.

The "liquid drop model" has been further refined by Myers and Swiatecki[41] to the "droplet model" including a term in A$^{1/3}$ as well as cross terms of A$^{1/3}$ with the asymmetry $(N - Z)/A$. As further reviewed by Hasse[42] this model recognizes effects of finite nuclear compressibility and the modified nuclear density just below the surface. In view of the conspicuous shell effects discussed below, it has rather too many free parameters, but its detailed description of the surface is of great interest in discussing deviations from spherical symmetry, including the strong distortions constituting the barrier against fission.

One of the most important uses of Eq.(7) is to predict the *β-stability line*. Because of the alternating term influencing even A values, this concept is normally applied to odd A only, and for the chemical reasons outlined above, we are usually asking the question what odd A value is stable for a given Z. Actually, the odd Z values 43 and 61 have no β-stable nuclei at all, and argon (Z = 18) and cerium (Z =

58) have no β-stable nuclei with odd A. These four exceptions are intimately connected with the closed neutron shells N = 50, 82, 20 and 82 discussed below. We may estimate the most stable Z, keeping A constant, by differentiating the sum of the asymmetry and Coulomb terms with respect to Z, finding the minimum energy for

$$Z = \tfrac{1}{2} A/(1 + 0.0075 \, A^{2/3}) \tag{8}$$

We may give four numerical examples involving A being cube-numbers:

$$
\begin{array}{lllll}
A = 64 & 125 & 216 & 343 & \\
Z = 28.57 & 52.63 & 85.04 & 125.41 &
\end{array}
\tag{9}
$$

The three first values are in excellent agreement with the observed positions of the β-stability line for a given Z-value. Actually, this curve is situated at the odd A value, if only one odd A is β-stable. In the cases (like chlorine 35 and 37) where two odd A are β-stable, their (even) average value is indicated. Thus, the β-stability line passes at A = 64 for Z = 29; at A = 125 for Z = 52; and is at equal distances from 216 by being A = 215 for Z = 85 and 217 for Z = 86 as seen from the Isotope Tables[32]. Obviously, the extrapolation to unexplored areas (like Z around 126) is strongly dependent on the exactly parabolic dependence of the asymmetry term, as well as on the ratio 0.0075 in Eq.(8) between the coefficients to the Coulomb and to the asymmetry terms in Eq.(7).

It is perfectly clear that some of the nuclear properties cannot be explained by the analytical functions of Z and A in the "liquid drop" model. Maria Goeppert Mayer[43] pointed out in 1948 that nuclei with either Z or N = (A − Z), or both, selected among the set 8, 20, 28, 50, 82 or 126 show several signs of specific stability. Thus, the number of such β-stable isotopes is conspicuously high; they tend to be much more frequent (both as elements, and as isotopic abundances); the very short-lived α-emitters occur just after lead 208 (Z = 82, N = 126) and the last non-radioactive isotope bismuth 209, but a characteristic set of α-emitting isotopes of samarium, gadolinium and dysprosium have N = 84 (or slightly above). The half-life 10^8 years of samarium 146 (between $2 \cdot 10^{15}$ years for neodymium 144 and 97 years for gadolinium 148) is too short for occuring in minerals, whereas 3 percent samarium 144 (with N = 82) is found to be stable. 15 percent of the samarium has A = 147 with α-half-life 10^{11} years.

It is difficult to believe in a strong analogy between the K-values for the electronic groundstates of monatomic entities in Eq.(4) and the preferred Z and N values in nuclei. A major difficulty was analyzed by Maria Goeppert Mayer[44] in 1950 that the non-relativistic Schrödinger equation does not at all predict the predominant closed shells 50, 82 and 126, but would predict 40, 70, 112 in disagreement with experience. There is no detectable effect at N = 40, and a certain stability around zirconium seems to be explained rather by N = 50. Already Hund[45] studied the order of one-particle levels in central fields U(r) of highly varying form, and the obvious first-order approximation expected is

$$1\,s \ll 1\,p \ll 1\,d < 2\,s \ll 1\,f < 2\,p \ll 1\,g < 2\,d < 3\,s \ll 1\,h < 2\,f < 3\,p \ll \ll 1\,i < 2\,g < 3\,d < 4\,s \ll \dots \tag{10}$$

where the double inequality signs refer to the energy jumps in a three-dimensional harmonic oscillator (U proportional to r^2). Each set of degenerate eigen-values correspond to the same parity, and are equidistant (like in the familiar one-dimensional model of the vibration of a diatomic molecule). If the Pauli exclusion principle allows two particles per eigen-value, the double inequality signs correspond to 2, 8, 20, 40, 70, 112, 168, ... and in general $(k + 2)\,[(k + 2)^2 - 1]/3$. It is noted that the "principal quantum number" n (occuring for a central field inversely proportional to r) is not relevant in nuclei, and the lowest energy for positive l is denoted 1 p, 1 d, 1 f, ... so $(n - 1)$ indicates the number of radial nodes for finite (but positive) r values, which would be $(n - 1 - 1)$ for nl in monatomic entities. It is more regrettable that nuclear physicists call $l = 6, 7, 8, 9, \dots$ i, j, k, l, ... and not i, k, l, m, ... The simple inequality signs in Eq. (10) indicate the removal of degeneracies when the U proportional to r^2 is modified in direction of a "squared wall" with U constant for $r < r_0$ and exceedingly high for $r > r_0$. Such a modification makes an additional radial node more objectionable than two angular nodes [8] and has the opposite effect of going from K = 1 to higher values in a monatomic entity. The most interesting consequences of the nucleonic shell model [44] are I (and the magnetic moments) of nuclei having either Z or N one unit *below* or *above* the closed-shell values (exactly like the alkaline-metal atoms were important for Rydberg). It turns out that one has to introduce spin-orbit coupling for positive l values (like one has to do, when studying X-ray or photo-electron spectra [27,46] of inner shells of atoms) but in nuclei, the *higher* $j = (1 + \frac{1}{2})$ is roughly 2 MeV *more stable* than the lower $j = (1 - \frac{1}{2})$. It remains an enigmatic question why the spin-orbit coupling has the opposite sign in nuclei compared with many-electron atoms, and it has even been suggested [47] that it might be explained by the interactions between the quarks (three in each nucleon) discussed later. The major effect of spin-orbit coupling is to split the $(2l + 1)$ degenerate eigen-values belonging to a high l-value immediately following a double inequality sign in Eq. (10) into $(l + 1)$ degenerate eigen-values having $j = (l + \frac{1}{2})$ below a new double inequality sign, whereas the rest having $j = (l - \frac{1}{2})$ remain after the new double inequality sign:

$$1\,s_{1/2} \ll 1\,p_{3/2} \ll 1\,p_{1/2} < 1\,d_{5/2} \ll 1\,d_{3/2} < 2\,s_{1/2} < 1\,f_{7/2} \ll 1\,f_{5/2} < 2\,p_{3/2} < < 2\,p_{1/2} < 1\,g_{9/2} \ll 1\,g_{7/2} < 2\,d_{5/2} < 2\,d_{3/2} < 3\,s_{1/2} < 1\,h_{11/2} \ll \dots \tag{11}$$

now occuring for 2, 6, 14, 28, 50, 82, 126, 184, 258, ... and in general $(k + 1)\,[(k + 1)^2 + 5]/3$. With exception of the neglected values 8 and 20, the spin-orbit coupling in Eq. (11) is a much better expression than the non-relativistic Eq. (10). It is noted that all the eigen-states between two double inequality signs in Eq. (11) (with exception of the last set before the next double inequality sign) have the same parity, and are ordered according to decreasing j. Klinkenberg [48] investigated the evidence for ordering of nlj values in nuclei further, but today, there remains no reason to believe in a universally valid "Aufbauprinzip" mainly because of the strong deviations from spherical shape in the regions between the closed-shell situations (to be discussed briefly below). A specific problem is the minor difference between the behaviour of protons and neutrons. Whereas $U(r)$ for neutrons is zero outside the nucleus, and roughly has the shape of a beaker-glass (a rounded-off square-well) below this horizontal line, there is an additional contribution of $U(r) = + Z/r$ outside the nucleus, and a parabolic maximum inside (like the bottom of a wine-bottle) for protons. This modified central field discourage low l-values relative to the high l-value concentrating its proton density close to the nuclear surface. The most general effect is filling a given neutron shell earlier than the proton shell with the same nlj as indicated by the asymmetry term in Eq.(7) establishing an equilibrium with the Coulomb term in Eq.(8). The classical Coulomb term is $3 Z^2/5 R$ for a constant charge distribution inside a sphere with radius R, but quantum mechanics replaces Z^2 by $Z(Z-1)$ in this expression, and it may be [38] that correlation effects further decrease the Coulomb term. Striking differences occur close to the 82-shell. The (in part radioactive) isotopes xenon 135, barium 137, cerium 139, neodymium 141 and samarium 143 (having N = 81) all have groundstates [32] with $I = \frac{3}{2}$ and even parity, whereas the two stable thallium (Z = 81) isotopes 203 and 205 have $I = \frac{1}{2}$ and even parity. The latter fact can be explained by $3 s_{1/2}$ having moved up to touch the last double inequality sign of Eq.(11) but the $I = \frac{3}{2}$ for N = 81 clearly shows that $1 h_{11/2}$ is filled, and the last neutron may be in $2 d_{3/2}$. On the other hand, bismuth 209 (Z = 83) has an odd groundstate with $I = \frac{9}{2}$ confirming $1 h_{9/2}$. This is not the case for the isotopes (with N = 83) barium 139, cerium 141, neodymium 143, samarium 145 and gadolinium 147 having odd groundstates[32] with $I = \frac{7}{2}$ as if $2 f_{7/2}$ is more stable for neutrons than $1 h_{9/2}$. The second-order effect of interprotonic repulsion becomes progressively more important for high Z values. There are good reasons[49−51] to believe that Z = 114 and 164 are more pronounced closed shells than Z = 126 and 184. The origin of this shift is that the six eigen-values $2 f_{5/2}, 3 p_{3/2}$ and $3 p_{1/2}$ are pushed up to higher energy by the particularly strong interprotonic repulsion for low l values, leaving a distinct gap at 114 (though $1 i_{13/2}$ is filled before), and by the same token, the ten eigen-values $2 g_{7/2}, 3 d_{5/2}, 3 d_{3/2}$ and $4 s_{1/2}$ are split off by a large gap at 164. This additional stability (and increased barrier height for fission) has great importance for the hope of making and detecting nuclei of translawrencium elements[8,16].

In many ways, it is entirely unexpected that there is an (even imperfect) analogy between the closed-shell K values in monatomic entities and Z and N singularities in nuclei. As pointed out by Gombas and Gaspar, the total binding energy of Z electrons to the nucleus of a neutral atom with Z between 5 and 90 is, within very narrow limits, $Z^{2.4}$ rydberg (to be compared with the rest-mass energy of an electron $m_0 c^2 = 2(137.0360)^2$ rydberg) whereas, within 10 percent on each side, the binding energy of all stable nuclei with A above 15 is A times 8 MeV (or 7.5 millimassunits). However, a much more profound distinction is that the two strongest bound electrons (in the 1 s orbital) each have a binding energy almost Z^2 rydberg. The ratio between the one-shot ionization energy [27] of 1 s and of the next-most strongest bound orbital 2 s decreases from 18 in neon to 5 in the heaviest elements (it will never reach the hydrogenic value 4 because of second-order[8,21] relativistic effects). On the other hand, the continuum states starting with the groundstate of M^+ occur within 10^{-5}

of the groundstate electronic energy of the heaviest neutral atoms. This may be compared with the binding energy Eq.(7) normally having the continuum states (corresponding to loss of a neutron or a proton) starting somewhere between 98 and 99.8 percent of the groundstate binding energy for A between 50 and 250. In this connection, it is interesting that Myers and Swiatecki[52] carefully evaluated the deviations of the total experimental binding energy from the "liquid drop" model Eq.(7). Obviously, they show slightly different values for differing Z combined with the same N (or for differing N for a given Z) but it is generally true that the negative deviations (less binding energy than expected) are less than 3 MeV in the middle between the closed shells, and the positive deviations are up to 6 MeV for $Z = 28$ and 50 and up to 12 MeV for $Z = 82$ (which are enhanced by the almost coinciding $N = 126$). By the same token, the positive deviations[52] approach 5 MeV for $N = 28$ and 50, but 8 MeV for $N = 82$, and 12 MeV for $N = 126$. Hence, the order of magnitude of the closed-shell oscillations of the total binding energy of a nucleus (from protons and neutrons) is 1 percent. It is very difficult to define this concept for monatomic entities, but it is likely that the total closed-shell stabilization for $K = 36$ is less than 0.1 percent of the total electronic binding energy of the krypton atom, and for $K = 86$ less than 10^{-4} of the total electronic energy of radon. This statement of proportionally larger closed-shell effects in nuclei does not at all have the corollary that the total wave-functions are closer to anti-symmetrized Slater determinants in nuclei than in monatomic entities. This would be neglecting the profound difference that electronic systems are "aristocratic" in the sense of having a definite low number of electrons exceedingly stabilized by a central field having the nucleus at origo, whereas nuclei are "democratic" creating their own, roughly constant, central field of attraction. Under these circumstances, it is absolutely excluded[38] that the protons and the neutrons form well-defined configurations in the strict sense[3] of the one-nucleon functions in Eq.(11). On the other hand, the many-electron atoms are also rather far from having anti-symmetrized Slater determinants as total wave-functions. The correlation energy[8] has the order of magnitude 0.7 eV times $Z^{1.2}$ (and hence proportional to the square-root of the total energy) and is larger than the first ionization energy of all atoms having Z above 11 (sodium).

All nuclei with even Z and even N (with a few exceptions situated very far from the β-stability line) have the totally symmetric groundstate with even parity and I zero. Fig. 1 gives the energy[32] of the first excited state (if below 1.7 MeV) which then has *always* $I = 2$ and even parity. At this particular point, the closed-shell nuclei are somewhat exceptional. Carbon 12, silicon 28 and calcium 48 have alright the first excited state (even 2) but at 4.439, 1.779 and 3.832 MeV, respectively. ($N = 50$) has the effect that the first excited state of zirconium 90 is (even 0) at 1.761 MeV and the next (even 2) at 2.186 MeV. However, this seems to be specific for $Z = 40$, since krypton 86, strontium 88, molybdenum 92 and ruthenium 94 have (even 2) as first excited state, at 1.565, 1.836, 1.509 and 1.428 MeV, respectively. Oxygen 16 and calcium 40 share the privilege of having three first excited states (even 0, odd 3 and even 2) at high energies, 6.049, 6.130 and 6.919 MeV in the former, and 3.352, 3.736 and 3.904 MeV in the latter case. The situation is much more peculiar in lead 208 having odd parity of the three first states at 2.614, 3.198 and 3.475 MeV having $I = 3, 5$ and 4.

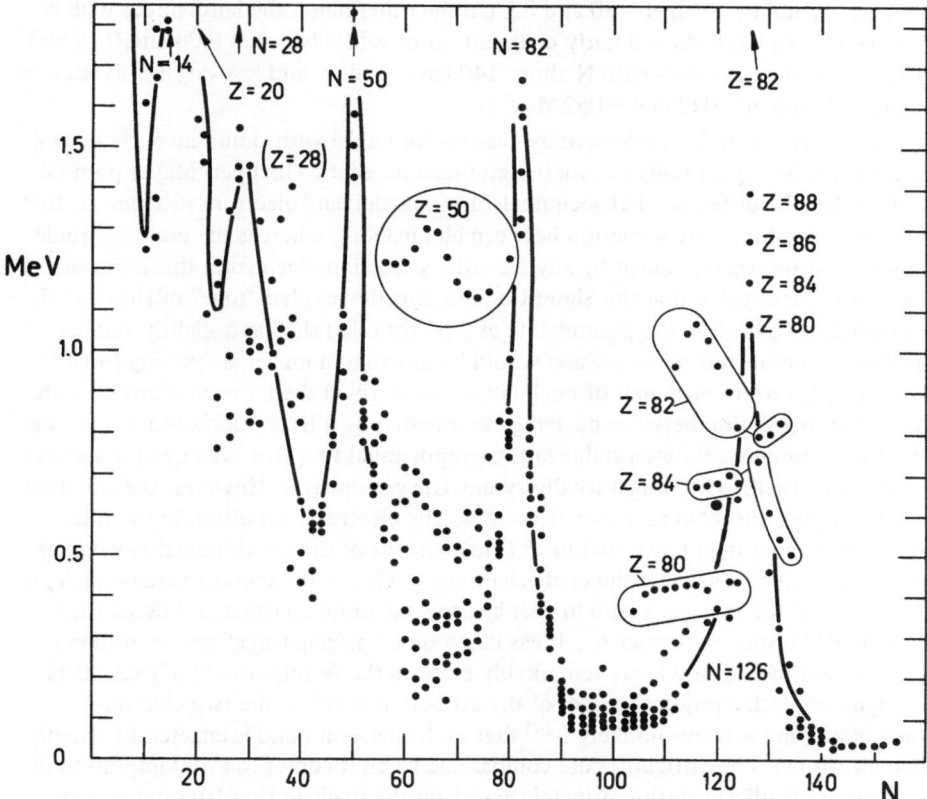

Fig. 1. The energy (in MeV) of the first excited state (always I = 2, if below 1.7 MeV) of nuclei having simultaneously even Z and even N. The points for special Z values are surrounded by round frames. Arrows pointing above 1.8 MeV are given for a few closed-shell cases.

The (even 2) energy is plotted as a function of N in Fig. 1. This leaves an inevitable dependence on Z. For instance, all the tin isotopes (Z = 50) have their first excited state in the 1.1 to 1.3 MeV range. It is difficult to study tin 132 (β-active with half-life 40 sec.) but its first excited state [32] at 4.04 MeV seems to have I = 3 and odd parity, like in lead 208. Figure 1 provides one of the most clear-cut arguments for the nucleonic shell model. It may be compared with another pragmatic plot, one of the precursors for the Periodic Table, where the molar volume of the solid elements show maxima at the alkaline-metals. Contrary to a plot of the first ionization energy of the neutral atoms, the volume plot shows rather symmetric peaks, the noble gases having long internuclear distances in the solidified state. The fairly symmetric character of the peaks on Fig. 1 is perhaps best seen for the frames containing points for Z = 80 and 84 somewhat below points for the closed-shell Z = 82. If both N and Z are reasonably far from the closed-shell values, the excitation energy generally decreases smoothly as a function of increasing Z for constant N. We note a broad distribution of excitation energies in the two minima between N =

28 and 50, and between N = 50 and 82. On the other hand, the lanthanides with N between 90 and 110 show a fairly coherent set of values between 0.08 and 0.15 MeV. The transradium isotopes with N above 140 have smaller, and less varying, excitation energies between 0.042 and 0.052 MeV.

There are several complementary reasons for nuclei with simultaneously even Z and even N having a totally symmetric groundstate. Maria Goeppert Mayer pointed out[44] that the difference between nucleons in nuclei, and electrons in atoms, is that we have a predominant attraction between all nucleons, whereas the interelectronic repulsion is not accompanied by any essential attraction. Seen from this point of view, it is conceivable that the Slater-Condon-Shortley explanation[2] of Hund's rule of highest possible S in the groundstate of a partly filled shell as a slightly smaller amount of interelectronic repulsion would be inverted in nuclei, a "pairing-force" favouring I zero for each pair of nucleons in partly filled shell. Calculations[44] with a contact interaction between nucleons represented by a Kronecker δ function in the total wave-function shows a stabilization proportional to q (for even q) of q nucleons in the same shell, exhibiting a totally symmetric groundstate. However, the situation is not simple either[4] in monatomic entities. The electronic repulsion in the filled shell d^{10} is more than twice that in d^6 (independent of the minor question whether S = 2 or 0), in part because the coefficient $q(q-1)/2$ to the average parameter A_* is 45 and 21 in the two cases, and in part because the more contracted d radial functions in d^{10} produce a higher A_*. Ideas based on a "pairing-force" and vanishing I for even numbers Z and N are remarkably close to the feelings of organic chemists, though they lack a physical model of the attraction between the two electrons in each bond (and it seems unlikely[8,24] that each chemical bond is enacted by exactly two electrons). The difficulties are comparable when it comes to avoid implosion of the system; Pauli's exclusion principle acting on the shells in Eq.(10) (and perhaps a "hard-core" repulsion between nucleons at short distances) is needed for the nuclear physicist, and rather sophisticated considerations of the electronic kinetic energy are needed for the organic chemist.

The first excited states are expected in the MeV region, using the nucleonic shell model. There is a very large number of I values feasible, and it would seem that the extended tables[32] are still quite incomplete. Much like the first excited J-levels of a noble gas atom belong to the configuration $(np)^5 (n+1 s)^1$ where an electron has been moved from the loosest bound, filled shell to the lowest empty orbital, one expect the first (odd) excited I-levels of lead 208 to be due to excitation of one of the 126 neutrons from $3 p_{1/2}$ to the low-lying empty $1 i_{11/2}$ (or another even orbital) or of one of the 82 protons from $3 s_{1/2}$ to $1 h_{9/2}$. The I values of these four excited levels are 5, 6, 4 and 5 not including the experimentally lowest I = 3. It is very difficult to have a clear idea of the (perhaps catastrophic) effects of configuration interaction on such excited levels. The β-active lead 209 has an even groundstate with I = $\frac{9}{2}$ suggesting the spin-orbit coupling to be sufficiently strong to make $2 g_{9/2}$ more stable for N = 127 than $1 i_{11/2}$ (of which the counterpart $1 i_{13/2}$ has been filled long before N = 126) again providing I = 4 and 5 by excitation of $3 p_{1/2}$. Actually, the odd groundstate of lead 207 (N = 125) has I = $\frac{1}{2}$ followed at 0.570 MeV by an odd level with I = $\frac{5}{2}$ looking like $2 f_{5/2}$. The excitations $2 f_{5/2} \rightarrow 2 g_{9/2}$ (or $1 i_{11/2}$) *can* produce an odd I = 3.

Ironically enough, Fig. 1 providing one of the most irrefutable arguments for the nucleonic shell model, also contains the basis for an alternative explanation of the excited I-values. As reviewed by Aage Bohr and Ben Mottelson [38] there are convincing reasons to believe that nuclei not too close to nucleonic Z or N closed shells are far removed from spherical shape. It was originally felt that the delicate balance between volume and surface effects in Eq.(7) would discourage strong deviations from the spherical condition minimizing the surface of a drop with constant density. However, a large amount of experimental evidence is in favour of groundstates normally being *spheroidal*. The general ellipsoidal case with three differing principal axes is not observed, but two of the axes are equivalent and different from the third axis. The spheroid can be *prolate* (elongated with the third axis longest) or *oblate* (with the third axis shortest). Such systems show *rotational spectra* (like diatomic molecules) with energies proportional to I (I + 1). Systems with totally symmetric groundstate show only the even I values. Actually, many of the nuclei in Fig. 1 with low excitation energy have four excited even levels [32] with I = 2, 4, 6 and 8 showing excitation energies in close agreement with the relation 6 : 20 : 42 : 72 expected. The moments of inertia of a rigidly rotating nucleus (with the volume well established from nuclear physics in general) do not agree with the observed energies, but have to be divided by a factor 2 to 3. This quantity represents the effective moment of inertia around a rotational axis perpendicular on the nuclear symmetry axis. There has been a large amount of recent work on interesting (and frequently dramatic) changes of effective moments of inertia for high I values, but it falls outside the scope of our discussion of groundstates. Some of the abrupt changes seem related to crossing of two rotational bands with different origin, much like strong perturbations between two series in atomic line spectra.

The separation of the $(j + \frac{1}{2})$ eigenvalues belonging to a definite (nlj) in spherical symmetry, when a strong perturbation of linear symmetry is introduced, was first treated by Nilsson [53] in 1955. The theory of coupling such ω-values with the rotational spectra of nuclei with odd A is very satisfactory [38,53,54] and shows features analogous to the rotational spectra of diatomic molecules having electronic states with positive Ω. Some apparent deviations from Eq.(11) may be due to such spheroidal behaviour. For instance, oxygen 17 (N = 9) with one expected 1 $d_{5/2}$ neutron has alright an even groundstate with I = $\frac{5}{2}$, and the first excited state (even I = $\frac{1}{2}$) conceivably due to 2 $s_{1/2}$ occurs at 0.871 MeV. However, the groundstate of fluorine 19 (Z = 9) is (even I = $\frac{1}{2}$) followed already at 0.1099 MeV by (odd I = $\frac{1}{2}$) and at 0.197 MeV by (even I = $\frac{5}{2}$). Ragnarsson, Nilsson and Sheline [54] recently wrote a review on shell structure in nuclei. These authors point out that the lowest rotational band (starting above the groundstate) of any nucleus with even Z and N occurs in radium 224, already at 0.216 MeV.

Between the two extremes of the spheroidal liquid drop (and its rotational spectra) and of the nucleonic shell model, one might still imagine a niche for the presence of α-particles. Much like the helium atom has its first excited level (S = J = 1, L = 0) at higher energy than any other neutral atom, the first excited I-level reported [32] of helium 4 occurs at 20.1 MeV (it may be noted from Eq.(6) that the energy needed to knock off a neutron is 19.80 MeV) and is totally symmetric. For the discussion below of the possible structure of baryons, it is very important to analyze the question what would be the observable consequences of the oxygen 16 being a regular tetrahedron constructed from four helium 4 nuclei. This is a very annoying subject for chemists. For instance, there is general agreement that the *rotational groundstate* (J zero) of a diamagnetic diatomic molecule (such as HCl) cannot *show* an electric dipole moment. However, since the heating bringing the molecule in a rotational state with positive J suffices to detect the electric dipole moment with external fields, the large majority of chemists believe that the groundstate contains a "meta-physical" dipole moment (which can be developed like a photographic plate by the moderate heating needed to populate rotationally excited states). In nuclei, a related quantity is the electric quadrupole moment Q. It can be argued [31] that the minimum quantum-mechanical uncertainty allowed for fixing the linear axis of the quadrupole moment in an external field multiplies the "genuine" value Q_0 by a factor (always below one) $I(2I-1)/(I+1)(2I+3)$ which is seen to cancel the observable moment for $I = \frac{1}{2}$ and zero. At a meeting in Villeurbanne 21. January 1980, Dr. Duval was so kind as to draw my attention to the similarity between the Q_0 potentially present in totally symmetric groundstates and its experimental manifestation for high I values (like 2 in Fig. 1) and the Jahn-Teller effect acting on degenerate eigen-states of electronic systems containing several nuclei. Anyhow, if a technique for an instan-taneous [8] determination of the internucleonic distances could be devised, a spherical system would show less dispersion than a "genuine" Q_0 of a spheroid. Coming back to oxygen 16, Robson [55] investigated the highly excited states, and interpret the (odd I = 3) as one of the rotational states of a regular tetrahedron. Though it is quite general in quantum mechanics that certain properties only become observable after excitation (a significant example are one-electron energies [27,46] in many-electron systems) the question of the tetrahedral rotational spectrum of the oxygen 16 nucleus invites the profound query whether *we* produce the properties by exciting the system, which were not at all present in the well-isolated groundstate.

3   Quarks in Baryons and Mesons, their Flavours and Colours, and the Scarcity of Unsaturated Quarks

Much attention became directed to the question of inner constituents of the nucleons by the paper "Structure of the proton" published by Feynman [56] in 1974. At that time, high-energy collisions between electrons confirmed the simple hypothesis that they are "point-shaped" in the sense of following Coulomb's law down to arbitrarily small distances. By the way, this does not prevent that the electron has a characteri-stic diameter = $e^2/m_0 c^2 = 2.818$ fm which is the classical expression for a charged soap-bubble having the electronic $m_0 c^2$ as electrostatic energy. In sufficiently high field strength [21], the vacuo is capable of providing pairs of electrons and positrons, and in this sense, it is the difference between the number of electrons and the number of positrons, which is conserved under dramatic conditions. However, in 1974, it was apparent that high-energy collisions between protons, or between protons and elec-trons, clearly suggested the presence of a small number of *partons* in the protons, carrying their own momenta, and having some of the aspects of nuclei detected in atoms by Rutherford scattering of α-particles. As described in the book[57] by Close: "An Introduction to Quarks and Partons", it was soon realized that, at least, a large proportion of these partons are identical with the *quarks* invented by Gell-Mann[58] with the main purpose of classifying correctly the eight lowest baryons (and their eight anti-particles) having $J = \frac{1}{2}$ (the symbol I is used for iso-spin, but J means the same as I for nuclei) with the rest-mass energies given in MeV, as customary in high-energy physics: proton (938.28), neutron (939.57), Λ^0 (1115.6), Σ^+ (1189.4), Σ^0 (1192.5), Σ^- (1197.3), Ξ^0 (1315) and Ξ^- (1321.3) as well as the ten baryons with $J = \frac{3}{2}$ stretching between the doubly charged Δ^{++} (1232) and Ω^- (1672). These 18 baryons can be constructed each of three quarks chosen between three categories *(flavours)*, and the corresponding 18 anti-baryons can be made from three anti-quarks. Glashow et al.[59] gave compelling theoretical reasons for a fourth flavour, and in particular, the "charmed"[60,124] mesons confirmed the utility (perhaps in addition to the necessity) of this fourth quark flavour. This subject is in a very rapid evolution, and has been reviewed by Mulvey[61] and by Marciano and Pagels[62]. It seems highly probable that five (and possibly six) flavours exist, with the names up, down, strange, charm and beauty. One of the most striking properties of the quarks is that their electric charge is a multiple of (e/3), *a-third* of the protonic charge:

$$u(2e/3) \quad d(-e/3) \quad s(-e/3) \quad c(2e/3) \quad b(-e/3) \tag{12}$$

Some of the baryons above should be in pure (or almost pure) quark configurations, such as the proton (uud), neutron (udd), Λ^0 (uds), Σ^+ (uus), Σ^0 (uds, using the orthogonal distribution of spins, relative to Λ^0), Σ^- (dds), Ξ^0 (uss) and Ξ^- (dss). Among the ($J = \frac{3}{2}$) baryons, the three extreme examples (which cannot provide $J = \frac{1}{2}$) are Δ^{++} (uuu), Δ^- (ddd) and Ω^- (sss). If quarks are genuine fermions, the latter examples can only satisfy the Pauli exclusion principle if each quark flavour is combined with another quantum number *colour*. The most important (and possibly only) values are three alternatives, which we give the Ostwald names "red", "yellow" and "blue" (though these allegories do not correspond exactly to more recent[63] colour technology) and a neutral element "grey". It must be noted in all fairness[57] that the three colours may be less "superficial" than a multiplication of otherwise identical systems with a new quantum number. Thus, a set of three linear equations allow different charges among the three colours combined with a given flavour, which would reproduce the same properties of the 8 and 10 baryons. One alternative compatible with these linear equations would be electric charges being an integer multiplying e, but Chanowitz[64] presented strong experimental arguments that this is not the case. Once fractional charges are admitted, the invariance of Eq. (12) with colour becomes a much more reasonable alternative.

The refreshing review "Quarks for Pedestrians" by Lipkin[65] and the remarks by the writer[66] keep the possibility open that quarks are not *essentially confined* three at a time in their baryons. Many authors[67,68] have defended an absolute confinement as a novel, unexpected but inevitable, feature of matter. In final analysis, this question shall only be decided by observations, but it is not a sufficient argument that present-day linear accelerators have not been able to provide collisions between protons and other particles (or between electrons and positrons colliding at GeV energies) liberating detectable quarks. Within the first second after the Big Bang 10^{10} years ago[35] quarks may have existed, among which a small fraction has not had the opportunity to recombine. It is likely that gravitational singularities *black holes* of the order of magnitude 10^{34} g (five solar masses), if not even much heavier, may tear rapidly passing proton apart, one or two of its quarks falling down in the black hole, though Jones[69] has given severe lower limits of abundance in cosmic rays. Another conceivable continuing source of quarks may be Hawking quantum evaporation[37] of small black holes (about 10^{15} g). Whereas neither photons nor other carriers of information can escape beyond the event horizon from the inside of a black hole in Einstein's relativistic theory, the quantum conditions at this surface produce emission of light comparable to the standard continuous spectrum of an opaque object ("black body") having an apparent temperature T inversely proportional to the mass of the black hole.

The only constructive evidence for unsaturated quarks are the experiments of Fairbank[70] of magnetic levitation of superconducting niobium balls, showing charges like $+0.34$ (or -0.66) by Millikan oil-drop type of measurements. Since roughly one unsaturated quark shows up on 10^{-4} g samples, one out of 10^{18} niobium nuclei (A = 93) or one out of 10^{20} nucleons are accompanied a quark or an anti-quark. These values are far higher than the higher limits (within the experimental uncertainty) reported for other materials, typically of the order one unsaturated quark per 10^{23} nucleons (this is still 6 quarks per gramme) though the detection without losses is a very difficult problem in ignorance of the atomic weights involved[66,71]. As analyzed by Orear[72] the niobium experiments indicate a certain mobility of unsaturated quarks under relatively mild handling of the samples, and it is conceivable[66] that positively charged quarks are combined with electrons and not with nuclei. De Rújula et al.[73] discussed why even positively charged quarks may show an enormous affinity for nuclei, and also suggested that there may occur "quark mines" like there are gold mines, due to the long-term geochemical separation processes[66]. Yock[74] has found three particles in cosmic radiation, of which the absolute values of the electric charge are 0.70, 0.68 and 0.42, and of which the atomic weights are above 4; 4 and 20, respectively. Their half-lifes are far longer than 10^{-8} s. On the other hand, a search[75] for oxygen isotopes with possibly (fractional) atomic weights in the interval 20 to 54 (using a tandem accelerator as an ultrasensitive mass spectrometer) yielded upper limits in the 10^{-18} to 10^{-16} range. It would be highly interesting to try to get limits around 10^{-20} for a few heavier elements.

The general consensus is that the "colour threshold" is certainly not passed by any particles lighter than 5 GeV (or 5.42 units of atomic weight, in the following called amu). This is only a very loose lower limit for the uncombined quark restmass, since the most likely candidate for the lightest adduct of a quark is a coloured meson (much like the pion is only 0.15 amu and the doubly charmed mesons[60] start above 3 amu). In the Gell-Mann picture[56] mesons are adducts of a quark and an anti-quark, and rather different from baryons "consisting" of three quarks (or anti-baryons "consisting" of three quarks (or anti-baryons of 3 anti-quarks). The asymptotic properties of quark adducts[57,65] are rather independent of the numerical value of the large rest-mass M of the free quarks. It can be calculated to the first approximation, that for both u- and d-quarks, the binding energy of a nucleon is approximately $(3M - 1)$, where 1 represents the atomic weight of the final product. The binding energy of a *diquark* has the order of magnitude M, and hence, its rest-mass is comparable to that of a free quark. Since the adduct of a nucleon and an anti-quark may have a binding energy E_b such a species has the energy $(M + 1 - E_b)$. The diquark is unstable toward the formation of a nucleon and an anti-quark, if the latter quantity is lower than the rest-mass of the adduct of two quarks. It is rather difficult to predict such binding energies; the lightest mesons (the neutral and the charged pions) show the attraction $(2M - 0.15)$ between the quark and the anti-quark, of which the rest-masses have been cancelled almost as effectively as by their total annihilation.

Recently, Wagoner and Steigman [76] revised the previous estimates of M and the average concentration of *primordial quarks*. They demonstrated that within quite broad intervals of the interval of parameters characterizing the "Big Bang" model (e.g. the uncombined quarks froze out of equilibrium at a quark-hadron transition with kT somewhere between 200 and 400 MeV corresponding to T between 2200 and 4400 inferno, Gamov's colloquial name for the temperature 10^9 K) the (experimentally suggested) ratio 10^{-25} to 10^{-20} between the numbers of uncombined quarks and of nucleons indicate an atomic weight M between 15 and 30. If M = 20, the combination of three quarks to a nucleon releases 98.34 percent of the original rest-mass. The writer [77] recently discussed the tendency toward totally symmetric groundstates as a function of increasing R, the ratio between the binding energy and the rest-mass of the final products. The order of magnitude of R is 10^{-10} for molecules formed from atoms, 10^{-6} for heavy atoms relative to their nucleus and free electrons, 10^{-2} for nuclei compared with protons and neutrons, and now perhaps 50 for nucleons relative to u- and d-quarks.

The total symmetry of baryons and mesons (heaving the generic name *hadrons*) involves grey colour (to the extent that the only non-grey systems reported are Fairbank's niobium balls) and among the hadrons, the baryons are not strictly totally symmetric, because they are fermions. Most people believe that Nambu [78] is correct in assigning anti-colours to anti-quarks, say "green", "violet" and "orange". Then, a grey meson may be constructed from a red quark and a green anti-quark, and so on. The writer [77] argues that colour may be the same for quarks and anti-quarks, like the addition of the sub-script in symmetry types of a super-group [8] may leave invariant the multiplication table of the lower group (here the colours). If this argument is valid, the grey and the red, yellow and blue colours would be isomorphous with Klein's Vierergruppe (like the point-groups D_2, C_{2v} and C_{2h}) having "grey" as neutral element, and the product of two non-grey colours being the third non-grey colour. Hence, a red quark and a red anti-quark would form a grey adduct. At least, it is not a convincing counter-argument that a not particularly stable grey diquark may be formed from two red quarks. Comparison with monatomic entities [77] shows that the totally symetric (even ^1S) groundstate characterizing closed shells (its character as neutral element of Hund vector-coupling is of prime importance for the Periodic Table) also may occur as highly excited states, e.g. of one partly filled shell containing an even number of electrons. Further on, it is by no means certain that "exotic" baryons not consisting of exactly three quarks may not occur in the lower half of the interval between 1 and M amu. As first pointed our by DeRújula, Giles and Jaffe [73] a quark (or an anti-quark) may be bound to the extent of several tenths, if not several times, an amu to a nucleus, forming what these authors call a "quarkleus" (quarklei in plural). The binding energy $(2M - 0.15)$ amu of the pion clearly shows that we must expect exorbitant conditions of attraction between uncombined quarks. If all the known hadrons have low energy (compared with M) we may also be prepared for more than three colours [79] at really high energy. The order of magnitude [80] of the atomic weight expected for the neutral Z and the charged (W^+ and W^-) bosons in Weinberg and Salam's *unified theory* of electromagnetic and weak interactions (responsible for the non-conserved parity in β-radioactivity) is 70 to 100.

The name "grand unified theory" is applied to a simultaneous description of electromagnetic, weak and strong interactions (producing the attractions between nucleons in nuclei). Gell-Mann, Ramond and Slansky [81] discussed the group-theoretical structures compatible with the grey and the three other colours in such a situation. One of the most striking consequences of the concomitant (exceedingly weak) mixing of quark and lepton characteristics in the "elementary" particles is that the baryon conservation rule no longer is strictly obeyed. In such a case, all matter is metastable (and if it is neutral, it breaks up in photons and neutrinos without rest-mass, in the final analysis). The theoretical prediction of the half-life of the proton is only one to two orders of magnitude longer than the present-day higher experimental limit of 10^{30} years. Several groups are working on detecting this radical disintegration of the proton [82] looking at 10^6 kg samples of water reasonably well protected against cosmic radiation. It is of obvious interest to predict what the oxygen nuclei (providing 89 percent of the nucleons in water) do under such experiments, and it seems [83] that about half of the yield of positive muons (and of various mesons) would be observed, relative to pure hydrogen. It is not obvious what influence the incorporation of a proton or a neutron in a nucleus has on the rate of its (very slow) disintegration. At one side, there is 1 percent less exothermic energy available (and it is known from the steep dependence of log $t_{1/2}$ for α-decay on the energy [33] how important this effect can be) but at the other hand, there may be much more efficient channels available in the nuclei. The writer has the hunch that one should not underestimate the interactions between quarks "inside" different nucleons; the quarks are spatially much less isolated than e.g. nuclei in adjacent atoms, and it seems [73] that the coloured system containing 13 quarks have properties entirely different from the α-particle with 12 quarks behaving much more clearly like 4 distinct nucleons. Seen from the point of the chemist, the former system containing unsaturated quarks have many of the properties of an aromatic radical (compared with an aliphatic compounds) or of a metal compared with solid argon.

At one time, the writer [77] imagined that the disintegration of a neutron in a beryllium 9 nucleus would produce two ultrarapid α-particles. However, as Howard Georgi suggested, the energy transfer to the nuclear fragments may be more effective in heavier nuclei, such as bismuth. The disintegration of an "interior" nucleon producing fission of ^{209}Bi would yield about six times more kinetic energy of the fragments than the spontaneous fission (which has not yet been detected). The most interesting samples to investigate are old transparent minerals, showing tracks due to the dramatic reactions. A million year old crystal shows 1 event/g for each unit of the ratio between

10^{30} years and the half-life of the nucleon disintegration. It is not fool-hardy to believe that certain 10^8-year-old crystals may contain constituents with average $t_{1/2}$ as short as 10^{28} years, in which case 10^4 events/g have taken place. Whereas beryllium forms transparent minerals, it is much more difficult to find limpid crystals with the elements from rhenium ($Z = 75$) to bismuth ($Z = 83$) as major constituents, and it may be that one has to concentrate on heavier lanthanides and tungsten showing fission tracks induced by nucleon disintegration. The major problem is the background of cosmic radiation; many authors have argued that the dinosaurs may have disappeared subsequent to a sudden burst of γ-ray photons from a close supernova explosion, and very extended effects are expected[84] with a frequency of roughly once in 10^8 years.

The observable effects of the radical disintegration of a nucleon may be imitated by some of the consequences of transfer of unsaturated quarks. It was discussed in the previous review in this series[66] how the energy levels of an atom containing a negative quark for purely electrostatic, non-relativistic reasons were similar to those of an atom containing a negative muon. Besides the X-rays emitted by a negative quark cascading down to its 1 s orbital just outside the nucleus, much more energetic events would be the subsequent capture of the negative quark by the nucleus (having a definite half-life) or the penetration of the Coulomb barrier by a positive quark having a large affinity[73] to the nucleus. Heavy elements (such as uranium and thorium, and perhaps even bismuth) may show fission or spallation events following quark capture with an energy far superior to that of spontaneous or neutron-induced fission. Zweig[85] analyzed the question whether heavy quarks may catalyze reactions between nuclei (such as deuterons) being brought at very close distance. Such catalysis is observed[86] with negative muons. Zweig notes that a particle with charge ($- 4e/3$) (an exotic quark, or a di-anti-u-quark) would bring two deuterons within a distance of 10^{-12} cm allowing rapid fusion. Whereas the conservation of electric charge was thought[66] to liberate the quark again after the nuclear reaction, it seems now likely[73] that the quark would be locked up in the helium nucleus formed, and the reaction would not proceed as a one-string catalysis.

Besides the interest in reactions evolving more than 95 percent of the rest-mass as mobile energy (rather than 1 percent in fusion of hydrogen isotopes, or 0.1 percent by fission of heavy elements) the major interest for chemists in species containing unsaturated quarks is that they constitute a *finer grid in the Periodic Table* with apparent atomic numbers $(Z + \frac{1}{3})$ and/or $(Z + \frac{2}{3})$. The predicted chemistry has already been sketched in a few areas[66] and may turn out to be accessible to study with techniques[16,87] related to the treatment of a few atoms of the transnobelium elements. Chemists should not underestimate that the rest-masses of the 18 lightest baryons are readily rationalized[88] by construction from u-, d- and s-quarks, as well as a large number of meson energies[89]. Quarks are slowly becoming indispensable, much like neutrinos worked their way out of the dark forest, but the great difference between these two kinds of theoretically yearned entities is the vanishing rest-mass of the neutrino and the huge, unknown rest-masses of the species containing unsaturated quarks. It is perfectly clear from high-energy collisions that the quarks bound inside the nucleons have very low *effective* masses (now frequently called current masses). There is not general agreement on the numerical values; the u- and d-quarks are assumed by Franklin[88] to be close to 0.15 amu, whereas the ratios between effective quark masses derived by Dominguez[90] are compatible with the very low values[91] 7 MeV for u-, 12 MeV for d- and 220 MeV for s-quarks. It is an intricate problem whether the low effective masses correspond to highly increased kinetic energies after confinement in a definite small volume. If the rest-masses of the uncombined quarks are at all defined, they would not produce spectacular relativistic contributions to the energy of a baryon, whereas Franklin[88] argues that relativistic effects are quite important. Recently, King and Rohrlich[92] suggested that relativistic effects dependent on the relative positions and momenta of two quarks may yield a dynamic confinement of three quarks in a baryon. This confinement due to velocity-dependent interactions is not of the "dogmatic" type[57] and may help in understanding the scarcity of unsaturated quarks.

For the purpose of detecting monatomic entities containing an additional quark on the nucleus, it would be highly helpful to calculate accurate positions of the spectral lines[93]. As pointed out by Fairbank, one might then hope to use the laser techniques of detecting a single atom, what has been done in practice with caesium[94,95,125]. Besides the well-recognized fact that species containing unsaturated quarks are not obliged to have atomic weights close to an integer A, it is not certain that the only opportunity of detecting them is that u- and d-quarks are not unconditionally confined. Several authors[96−98] have drawn attention to the possibility of exotic quarks producing very long-lived hadrons, protected against decay by very strong selection rules.

4 Rishons or Other Subquarks, and the Generations of Leptons and Quarks

Quantum mechanics has to a large extent resolved an antinomy inherited from the discussions of Heraclit and Democrit. The wave-functions are continuous and extended in the former sense, but at the same time, the "indivisible" parts of Democrit have been replaced by normalization conditions, the numbers K, Z, N, . . . of electrons, protons, neutrons, . . . being cardinal numbers without any possible way of assigning ordinal numbers to the individual, indiscernible entities. This trend has been further accentuated by most particles having anti-particles, with exception of some bosons (such as the photon and the neutral pion, but not the α-particle).

The essentially point-shaped characteristics of quarks and leptons made it plausible[61,62] in 1979 that they represent the smallest constituents of matter. Whereas it would be disrupting quantum mechanics to accommodate fractional angular momenta other than those permitted for bosons and for fermions, there is no convincing proof that the electric charge e of the proton is the smallest conceivable. We bifurcate here in a new antinomy: either is further division of electric charges in quanta smaller than (e/3) feasible, but energetically highly expensive; or otherwise, we hit rock bottom of the ocean with the lower limit (e/3). In the autumn 1978, the writer suggested in a preprint that the conspicuous importance of binary classifications[99] made an extrapolation of Klein's Vierergruppe plausible, in such a way that $(2^k - 1)$

symmetry types analogous to quark colour, and one neutral element, establish a series of stratifications with the ratio R between binding energy and the rest-mass of the final products increasing dramatically with k. Thus, R might be 10^6 or 10^{10} for the seven *quips* conceivably constituting a quark, each having a charge being a small multiple of $(e/21)$. It is evident that this line of thought provides some support for the suspicion[67,68] that *we* induce the fine-grained structure by providing high energy, as we already discussed above the highly excited rotational states of ^{16}O nuclei suggesting tetrahedral shape. It would no longer be as appropriate to think of quarks as "constituents" of a baryon, as to consider quarks as a helpful device for describing high-energy reactions of hadrons. Obviously, it cannot be decided today whether such an infinite regression of sub-structures is incompatible with future experimentation, but the opposite point of view has got impressive support from two papers published by Harari[100] and Shupe[101] the 10. September 1979.

Before discussing the two (rather similar) proposals of Harari and Shupe in detail, it is useful to consider some arguments about *generations* making a classification perpendicular on the dividing line between quarks and leptons. The emphasis has moved at two spots. As long the effective mass of quarks moving inside hadrons was considered to increase dramatically along the series s, c, b, . . . after having been quite small for u and d, there was no clear-cut analogy with the leptons. The neutrino and the anti-neutrino having no rest-mass; and the positron and the electron having the lowest well-established positive rest-mass; is followed by an entirely analogous series: the muonic neutrino, the muonic anti-neutrino, the positive and negative muon. It remains a debatable question whether the muonic neutrino and anti-neutrino have rest-masses at all (though they are quite distinct from the electronic neutrino in their reactions at high energies) and it is definitely below 0.5 MeV, whereas the rest-mass 105.660 MeV of the charged muon is in striking contrast to the electron, when it is realized that their intrinsic magnetic moments are the same within 6.10^{-6}. Barut[102] argued that the mass ratio between the charged muon and the electron is $1 + 3 (137.036 . . .)/2$ obtained from the two solutions of a generalized Dirac equation. Whereas the muons still are lighter than the lightest mesons (i.e. the pions), a new set of leptons with the positive and negative τ particles (in the following called tauons) shown in 1978 to have rest-mass 1785 ± 5 MeV disqualified somewhat the name "lepton". Barut[103] suggested a recursive formula for the mass of the charged lepton relative to the electron mass m_0

$$m_n = m_{n-1} + (\tfrac{3}{2} \cdot 137.036 \, n^4) \, m_0 \tag{13}$$

where 1786.08 MeV is predicted for $n = 2$ and 10293.7 MeV for $n = 3$. Eq. (13) was derived from the Bohr-Sommerfeld quantization of a charge moving in the field of a magnetic moment, and seems to fall in the same category of heuristic results as the 1913 formula for the hydrogen atom. In the following, we call the electronic, muonic and tauonic generations β, μ and τ. The number of different neutrino species have very marked consequences for the concentration of deuterium and helium remaining after the "Big Bang". The conclusions[104−106] reported are, among others, that the muonic neutrino is likely to be lighter than 50 eV, and that if the tauonic neutrino is heavier than 50 eV, then it is heavier than 10 MeV and unstable towards decay to a lower neutrino and a positron-electron pair, with a half-life shorter than one day. What is perhaps more important for our purposes is that *less* than four generations of neutrinos lighter than 1 MeV and with half-life above 1 s is a necessary consequence[104] of cosmological observed parameters, leaving only the space for three known generations of leptons.

The earliest argument for classifying quarks and leptons in the same generations was the phantastic agreement (better than 10^{-19}) between the electric charges (with opposite sign) of a proton and an electron. A more sophisticated step is the grand unified theory of Georgi and Glashow[82,107,108] demanding 24 bosons. Of these, 12 acquire enormous rest-masses, 10^{23} to 10^{24} eV (10^{14} to 10^{15} amu) and 8 mediate the strong interaction and have moderate, if not vanishing, rest-masses. Among the 4 remaining bosons, a second instance of spontaneous symmetry breaking takes place, and three (W^+, W^- and Z^0) have rest-masses in the 10^{11} eV range, whereas the last boson is the genuinely zero-rest-mass photon. The forces transforming a quark into a lepton (with valuable information derived from the lower experimental limit 10^{30} years for the proton decay) are a fifth, *hyperweak interaction*, in addition to electromagnetic, weak, strong and gravitational interactions. The characteristic distance of the hyperweak interaction is 10^{-29} cm, as much smaller than a proton radius as a proton is smaller than a football.

It is quite striking that the muonic and tauonic leptons look like excited states of the β generation. It would now seem plausible to ascribe two quark flavours to each generation, one with charge ($+ 2e/3$) and the other with ($- e/3$). If we count the three quark colours combined with each flavour, each generation has 8 particles 8 anti-particles as members. It is remarkable that all ordinary matter (outside high-energy physics apparatus) consists of the first β generation, since electrons and the u- and d-quarks (providing protons and neutrons) is all what you need, neglecting an unidentified number of neutrinos and anti-neutrinos also belonging to the first generation. The second generation contains the muonic leptons, c- and s-quarks. The situation of "asymptotic freedom" (vanishing interactions between two quarks at a short distance) requires[109,110] at most 16 quark flavours (and hence 8 generations) below the grand unification mass. Frampton, Nandi and Scanio[111] estimate that six such generations occur, though only the three first have been observed until now.

Harari[100] and Shupe[101] consider mainly the β generation of quarks and leptons. Harari introduces two *rishons*, T (with the charge e/3) and V (neutral), and the anti-rishons $\overline{T}$ ($- e/3$) and $\overline{V}$ (also neutral). The neutrino has the formula (VVV) and the anti-neutrino $(\overline{VVV})$. The other, simplest, fermions also contain three rishons, such as the positron (TTT) or the electron $(\overline{TTT})$. Whereas the leptons have their three rishons identical, the u-quark (TTV) and the d-anti-quark (TVV) have two of their rishons different from the third. The red, yellow and blue colour of the quarks is ascribed to this degeneracy. It is noted that the neutral atom containing a nucleus characterized by Z and N, altogether "contains" ($6Z + 3N$) rishons and ($6Z + 6N$) anti-rishons. There must be an enormous activation barrier against disproportionation reactions such as 2 (TTV) → (TTV) + (TTT) capable of disintegrating the proton. Whereas the proton contains 6 rishons and 3 anti-rishons, it is noted that the triple structures never mix rishons and anti-rishons. It may finally be remarked that positive rest-mass seems to be connected with the presence of electric charge in T and $\overline{T}$

though the atomic weight of (TTT) of 10^4 to 10^5 times smaller than of the quarks (TTV) and $(\overline{\text{T}}\overline{\text{V}}\overline{\text{V}})$.

Shupe [101] uses the name *quips* (which may be a rather obvious association of ideas) for essentially the same two rishons and two anti-rishons. Besides a series of interesting arguments about the mechanism of interaction between two quarks, which would be a little outside the scope of this review, Shupe also asks the fascinating question whether a photon "consists" of $(\text{T}\overline{\text{T}})$ and hence needs to triplicate before forming e.g. a positron and an electron, or whether it always has three rishons and three anti-rishons, even at low energy (where dissociation to a neutrino and an anti-neutrino has not been observed). It is evident that rishons are very far from being "Lego bricks" and rather represent topological twists in the vacuo state. Hence, a neutral pair $(\text{V}\overline{\text{V}})$ would probably have no more meaning than to say that a car takes simultaneously a turn of the left and a turn to the right.

If rishons are accepted as indivisible, the smallest amount of electric charge is indeed $(e/3)$. Both Harari and Shupe agree that rishons are not genuine fermions, only triple adducts (and more complicated beings) are. One has to be quite careful with the necessary triality; we have the same problem as with two or four quarks which may form stable adducts. Besides 3 being the lowest odd number above 1 (once it is argued that a single rishon cannot occur isolated) it may be noted [77,112] that 3 and 0 are the only numbers x of objects having x relations of the type of mutual distances.

It is well-known [56,113] that the magnetic moments of the seven lightest baryons (typically within some 10 percent) can be calculated from a static model of quark magnetic moments. The deviations (far beyond experimental uncertainty) have much of the character of gyromagnetic factors g in intermediate coupling for electronic states of monatomic entities. Several authors [114,115] have pointed out the difficulty for rishon models that the additivity does not at all provide the observed magnetic moments. There are several possible solutions of this problem. If it has any sense to say the inter-rishon distance in a quark or in a lepton is below 10^{-20} cm, the higher-order relativistic effects may be divergent in a way precluding any known method of evaluating the magnetic moment. The general expectation seems to be that rishons do not carry proper rest-mass; all their energy is stored in intense local fields. The other way out of the dilemma may be what Shupe [101] has hinted that the three rishons may occupy the same geometrical point. Such behaviour would enforce the feeling that they are quantum numbers much more than they are building-stones, but it would not be entirely ludicrous for the lowest stratification of matter.

There has been three other, more recent, proposals of structure in quarks and leptons. Taylor [116] suggested a model of composite quarks and leptons, where the three constituents carry solely flavour, colour and ancestor (generation) attributes. One corollary is that in a given generation, the grey colour represents the leptons, and that quarks so to say are coloured leptons. This book-accounting might seem rather formal, though strong evidence is available [117] that the potential of interaction between a quark and an anti-quark is independent of their flavours. Casalbuoni and Gatto [118] performed a rather abstract treatment starting with a left-handed fermion with $(J = \frac{1}{2})$, the electric charge Q zero, and the difference $(B - L)$ between the baryon number and the lepton number equal to (-1). A set of three creation

operators (with differing colour indices) carrying $Q = \frac{1}{3}$ and $(B - L) = \frac{2}{3}$ produce
the neutrino, d-anti-quark, u-quark and positron when applied consecutively to the
basic fermion. A grey creation operator carrying $Q = -1$ and $(B - L) = 0$ produce
the corresponding anti-particles. One advantage is that this description readily provide unified groups such as SU (5) and O (10). Most recently, Królikowski [119] proposed *coloured preons* (the name "preons" for subquarks was first used by Pati and
Salam in 1975) of which δ (occuring in three colours) is a fermion with $J = \frac{1}{2}$ and χ
(also with three colours) is a boson with zero spin. The two simplest members of the
β generation is the neutrino "consisting" of one δ and one anti-χ, and the u-quark of
one anti-δ and one anti-χ. *Both* δ and χ carry the charge $(-e/3)$ and their anti-preons
$(+e/3)$. The electron "consists" of one δ and *two* χ and the d-quark of one anti-δ and
two χ. It is also mentioned [119] that the mediating weak bosons of Weinberg and
Salam [80] conceivably may have a related structure, W^- and W^+ of three χ or three anti-χ
and the neutral Z a linear combination of $\overline{\delta\delta}$ and $\overline{\chi\chi}$ states (the orthogonal combination
being a candidate for describing the photon). Since all the preons are electrically charged,
the neutral entities must necessarily contain the same number of preons and anti-
preons (whereas co-existence of rishons and anti-rishons are forbidden in the same
quark, though it may perhaps occur in the photon). There are many attractive features of Królikowski's description by "primordial quantum chromodynamics" of
the first generation, but it seems slightly awkward that the d-quark contains three
"subquarks" like an electron, but the u-quark only two like a neutrino. By the way,
the d-quark can be constructed by adding an electron and removing a neutrino from
the u-quark, or in a certain sense, by adding three χ. This result might throw some
light over the enigmatic question why a neutron is a tiny bit (2.5 electron rest-masses) heavier than a proton, the substitution of one u-quark by one d-quark having
the same effect as replacing a neutrino by an electron. Królikowski [119] insists that
preons are *very* tiny and have huge masses.

Elbaz and Meyer [120] have proposed a "bootstrap" topological approach to both
quarks and leptons, where the T and V rishons are vectors in a space having the
observable particles as scalars. Also the W and Z bosons can be included. These authors attempt to *derive* Pauli's exclusion principle for fermions from the properties
of rishons.

As recently mentioned [112] we are faced with two plausible alternatives at the
moment. We may have at least the 24 particles and 24 anti-particles (counting the
colours of the quarks) belonging to the β, μ and τ generations. This is rather many,
though their pattern is far more appealing than the wildly proliferating "elementary"
particles from the time before Gell-Mann proposed the quarks. The other alternative
is that we have returned, like a pendulum, to the opinion prevailing just before 1930,
when matter was supposed to consist exclusively of protons and electrons. It may be
sufficient to have two rishons (or two preons) and their anti-particles. The major difference is that rishons are charged *or* neutral, and that preons are fermions *or* bosons.

The enormous amount of sophisticated work and huge funds invested in high-
energy physics the last thirty years have produced a plethora of experimental data,
of which the categorization and rationalization has much in common with what hap-
pened to chemistry in the century between Boyle and Lavoisier. We may very well be
stuck with leptons and quarks for an indefinite length of time, in close analogy to the

elements as viewed by Lavoisier. On the other hand, the ideas of subquarks (rishons or preons) are perhaps at the same time as specifically wrong and as heuristically fruitful as Prout's hypothesis from 1815.

It is evident that the growing familiarity with the application of quantum mechanics to the structure of matter introduces a Pythagorean emphasis on numbers, and make our "common sense" ideas of constituents fade away. In particular, small identical systems are thoroughly *identical*. This is true already for molecules of (definite isotopic constitution and) reasonable size and for anything below. It is possible to analyze the question [121,122] whether quantum mechanics is only applicable to systems so small that they can be exactly reproduced (and not, for instance, to animals and coins). Accepting this sceptical attitude is perhaps not a cowardice when we contemplate the methodological dilemma, and it certainly removes a lot of vexatious paradoxes. Going down along the hierarchical strata from monatomic entities to nuclei to quarks, it is clear that the urge toward cardinality and non-individuality grows irresistible in direction of the subquarks. The discussion of unconditional confinement of quarks, such as the "calypsons" of Drell[67] may have been a preliminary preparation to recognize the intrinsic confinement of rishons (and presumably preons) which are unable to subsist alone, one at the time. This also has an historical analogy in Ampère suggesting that magnetic dipoles always are due to electric currents.

As for many other problems of microscopic properties of matter, astrophysics[34,35,37] may be extremely helpful, both by large-scale observations of the Universe, and by observing matter under conditions which are not yet available in the laboratory. Fechner and Joss [123] analyzed the conditions for neutron stars (which are nuclei with N close to 10^{57}) performing a phase transition to "quark stars" at a density slightly below 10^{15} g/cm^3 and a pressure in the range 10^{29} atm. or of 10^{35} erg/cm^3. Surface atoms of such quark stars may emit spectral lines red-shifted about a-third of their normal wave-number. If their mass is above 1.8 ± 0.2 solar masses, they contract irreversibly to Einstein-Schwarzschild black holes. Before this collapse, such objects have a radius of about 10 km.

The most spectacular lacuna in the subquark models is the lack of a semi-qualitative estimate of rest-masses. In the rishon model, (VVV) has no rest-mass, and the rest-mass of (TTT) is many thousand times smaller than of a quark "containing" a mixture of T and V rishons. In the preon model, the same increase by at least a factor 10^4 occurs when the δ in the electron is replaced by an anti-δ forming the d-quark (as if the simultaneous presence of an anti-preon made the system much heavier in contradistinction to an anti-quark in a meson). The repulsive relation between δ and χ is further illustrated by the anti-u-quark "containing" these two preons, whereas the neutrino and anti-neutrino have one of two types being an anti-preon. The conservative point of view (that subquarks are not needed) is supported by the absence of any experimental evidence for leptons and quarks not being point-shaped. It is generally agreed that hadrons (both baryons and mesons) are far more extended in space than quarks and leptons. This statement forms a striking contrast to the feelings of the chemist considering the density (slightly above 10^{14} g/cm^3) of nuclei exceptionally high. However, the reason why we cannot form a liquid of quarks with a density of, say, 10^{30} g/cm^3 (which might form a threatening black hole unless such germs[37] are dissipated by Hawking evaporation) is that the kinetic energy of quarks with an atomic weight[76] close to 20 would be exorbitant. The typical densities (of order 1 to 10 g/cm^3) of condensed matter are determined by the two facts that the electron has a very low rest-mass (compared with nuclei) producing a high kinetic energy by confinement, even in relatively large volumes[8] and that the ionization energy of all neutral atoms, molecules and non-metallic solids is situated[13,27] in an interval between 3 and 25 eV, demanding (via the virial theorem) very low kinetic energies of the loosest bound electrons. Hence, chemistry is about electronic structure, and the zippy electrons hide away the ponderous specks of quark agglutinations from our eyes.

References

1. Hund, F.: Linienspektren und Periodisches System der Elemente. Berlin, Julius Springer 1927
2. Condon, E. U., Shortley, G. H.: Theory of Atomic Spectra (2. Ed.) London, Cambridge University Press 1953
3. Jørgensen, C. K.: Orbitals in Atoms and Molecules. London, Academic Press 1962
4. Jørgensen, C. K.: Oxidation Numbers and Oxidation States. Berlin, Heidelberg, New York, Springer 1969
5. Jørgensen, C. K.: Adv. Quantum Chem. *11*, 51 (1978)
6. Moore, C. E.: Atomic Energy Levels, Nat. Bur. Stand. Circular no. 467, vols. I (1949), II (1952) and III (1958)
7. Martin, W. C., Zalubas, R., Hagan, L.: Atomic Energy Levels, the Rare-Earth Elements. NSRDS-NBS *60*, Washington, Nat. Bur. Standards 1978
8. Jørgensen, C. K.: Modern Aspects of Ligand Field Theory. Amsterdam, North-Holland 1971
9. Jørgensen, C. K.: Israeli J. Chem. *19*, 174 (1980)
10. Reisfeld, R., Jørgensen, C. K.: Lasers and Excited States of Rare Earths. Berlin, Heidelberg, New York, Springer 1977
11. Jørgensen, C. K.: Gmelins Handbuch d. anorgan. Chemie, Seltene Erden *39 B1*, 17 (1976)
12. Jørgensen, C. K.: Handbook on the Physics and Chemistry of Rare Earths *3*, 111. Amsterdam, North-Holland 1979
13. Jørgensen, C. K.: Structure and Bonding *22*, 49 (1975)
14. Campagna, M., Wertheim, G. K., Bucher, E.: Structure and Bonding *30*, 99 (1976)
15. Wallace, W. E., Sankar, S. G., Rao, V. U. S.: Structure and Bonding *33*, 1 (1977)
16. Jørgensen, C. K.: Lanthanides and Elements from Thorium to 184. London, Academic Press (in preparation)
17. Jørgensen, C. K.: J. chim. physique *76*, 630 (1979)
18. Keller, C.: The Chemistry of the Transuranium Elements Weinheim/Bergstr., Verlag Chemie 1971
19. Edelstein, N., Brown, D., Whittaker, B.: Inorg. Chem. *13*, 563 (1974)
20. Gruber, J. B., Morrey, J. R., Carter, R. G.: J. Chem. Phys. *71*, 3982 (1979)
21. Fricke, B.: Structure and Bonding *21*, 89 (1975)
22. Pyykkö, P.: Adv. Quantum Chem. *11*, 353 (1978)
23. Englman, R.: The Jahn-Teller Effect in Molecules and Crystals. London, Wiley 1972
24. Jørgensen, C. K.: Chimia *31*, 445 (1977)
25. Bourcet, J. C., et al.: Chem. Phys. Letters *61*, 23 (1979)
26. Reinen, D., Friebel, C.: Structure and Bonding *37*, 1 (1979)
27. Jørgensen, C. K.: Structure and Bonding *30*, 141 (1976)

28. Sutton, L. E. (ed.): Tables of Interatomic Distances and Configurations in Molecules and Ions. London, Chemical Society 1958 and 1965
29. Elton, L. R. B.: in Landolt-Börnstein, Group 1, Vol. 2 (Nuclear Radii). Berlin, Heidelberg, New York, Springer 1967
30. Griffith, J. S.: The Theory of Transition-metal Ions. London, Cambridge University Press 1961
31. Lucken, E. A. C.: Nuclear Quadrupole Coupling Constants. London, Academic Press 1969
32. Lederer, C. M., Shirley, V. S. (eds.): Table of Isotopes, 7. Ed. New York, Wiley 1978
33. Gallagher, C., Rasmussen, J. O.: J. Inorg. Nucl. Chem. *28*, 741 (1966)
34. Trimble, V.: Rev. Mod. Phys. *47*, 877 (1975)
35. Weinberg, S.: The First Three Minutes. New York, Bantam Books 1979
36. Unsöld, A.: Naturwissenschaften *63*, 443 (1976)
37. Lorenz, D., Reinhardt, M.: Naturwissenschaften *66*, 390 (1979)
38. Bohr, A., Mottelson, B. R.: Nuclear Structure. New York, Benjamin (Vol. 1) 1969 and (Vol. 2) 1975
39. Green, A. E. S., Engler, N. A.: Phys. Rev. *91*, 40 (1953)
40. Zeldes, N., Grill, A., Simievic, A.: Mat. Fys. Skrifter Dan. Vid. Selskab 3, no. 5 (1967)
41. Myers, W. D., Swiatecki, W. J.: Ann. Phys. (N.Y.) *55*, 395 (1969)
42. Hasse, R. W.: Annals Phys. (N.Y.) *68*, 377 (1971)
43. Mayer, M. G.: Phys. Rev. *74*, 235 (1948)
44. Mayer, M. G.: Phys. Rev. *78*, 16 and 22 (1950)
45. Hund, F.: Handbuch d. Physik (ed. S. Flügge) *36* (Atome II) 1 (1956)
46. Jørgensen, C. K., Berthou, H.: Mat. fys. Medd. Dan. Vid. Selskab *38*, no. 15 (1972)
47. Pirner, H. J.: Phys. Lett. *85B*, 190 (1979)
48. Klinkenberg, P. F. A.: Rev. Mod. Phys. *24*, 63 (1952)
49. Larsson, S. E., Leander, G., Ragnarsson, I., Randrup, J.: Physica Scripta *10A*, 65 (1974)
50. Nurmia, M.: Physica Scripta (Stockholm) *10A*, 77 (1974)
51. Kolb, D.: Z. Physik *A 280*, 143 (1977)
52. Myers, W. D., Swiatecki, W. J.: Nucl. Phys. *81*, 1 (1966)
53. Mottelson, B. R., Nilsson, S. G.: Phys. Rev. *99*, 1615 (1955)
54. Ragnarsson, I., Nilsson, S. G., Sheline, R. K.: Physics Reports (Phys. Lett. *C*) *45*, 1 (1978)
55. Robson, D.: Phys. Rev. Lett. *42*, 876 (1979)
56. Feynman, R. P.: Science *183*, 601 (1974)
57. Close, F. E.: An Introduction to Quarks and Partons. London, Academic Press 1979
58. Gell-Mann, M.: Phys. Rev. Letters *8*, 214 (1964)
59. Glashow, S. L., Iliopoulos, J., Maiani, L.: Phys. Rev. *D2*, 1285 (1970)
60. Richter, B.: Rev. Mod. Phys. *49*, 251 (1977)
61. Mulvey, J.: Nature *278*, 403 (1979)
62. Marciano, W., Pagels, H.: Nature *279*, 479 (1979)
63. Hoshino, M.: Amer. J. Phys. *47*, 573 (1979)
64. Chanowitz, M. S.: Phys. Rev. Lett. *44*, 59 (1980)
65. Lipkin, H. J.: Physics Reports (Phys. Lett. *C*) *8*, 173 (1973)
66. Jørgensen, C. K.: Structure and Bonding *34*, 19 (1978)
67. Drell, S. D.: Physics Today *31* (June 1978), 23 (1978)
68. Capra, F.: Amer. J. Phys. *47*, 11 (1979)
69. Jones, L. W.: Phys. Rev. *D17*, 1462 (1978)
70. LaRue, G. S., Fairbank, W. M., Phillips, J. D.: Phys. Rev. Lett. *42*, 142 and 1019 (1979)
71. Jones, L. W.: Rev. Mod. Phys. *49*, 717 (1977)
72. Orear, J.: Phys. Rev. *D20*, 1736 (1979)
73. De Rújula, A., Giles, R. C., Jaffe, R. L.: Phys. Rev. *D17*, 285 (1978)
74. Yock, P. C. M.: Phys. Rev. *D18*, 641 (1978)
75. Middleton, R. et al.: Phys. Rev. Lett. *43*, 429 (1979)
76. Wagoner, R. V., Steigman, G.: Phys. Rev. *D20*, 285 (1979)
77. Jørgensen, C. K.: Arch. Sciences (Genève) *32*, 201 (1979)
78. Nambu, Y.: Sci. Amer. *235* (Nov. 1976), 48 (1976)

79. Wilczek, F., Zee, A.: Phys. Rev. *D16*, 860 (1977)
80. Quigg, C.: Rev. Mod. Phys. *49*, 297 (1977)
81. Gell-Mann, M., Ramond, P., Slansky, R.: Rev. Mod. Phys. *50*, 721 (1978)
82. Robinson, A. L.: Science *206*, 670 (1979)
83. Sparrow, D. A.: Phys. Rev. Lett. *44*, 625 (1980)
84. Ruderman, M., Truran, J. W.: Nature *284*, 328 (1980)
85. Zweig, G.: Science *201*, 973 (1978)
86. Alvarez, L. W., et al.: Phys. Rev. *105*, 1127 (1957)
87. Hubert, S., Hussonnois, M., Guillaumont, R.: Structure and Bonding *34*, 1 (1978)
88. Franklin, J.: Phys. Rev. *D21*, 241 (1980)
89. Wills, J. G., Lichtenberg, D. B., Kiehl, J. T.: Phys. Rev. *D15*, 3358 (1977)
90. Dominguez, C. A.: Phys. Lett. *86B*, 171 (1979)
91. Y. J. Ng, S. H. H. Tye: Phys. Rev. Lett. *41*, 6 (1978)
92. King, M., Rohrlich, F.: Phys. Rev. Lett. *44*, 621 (1980)
93. Rau, A. R. P.: Phys. Rev. *A17*, 1721 (1978)
94. Robinson, A. L.: Science *199*, 1191 (1978)
95. Jørgensen, C. K.: Naturwissenschaften *65*, 484 (1978)
96. Poggio, E. C., Schnitzer, H. J.: Phys. Rev. *D15*, 1973 (1977)
97. Cahn, R. N.: Phys. Rev. Lett. *40*, 80 (1978)
98. Zee, A.: Phys. Lett. *84B*, 91 (1979)
99. Jørgensen, C. K.: Logique et Analyse (Louvain) *7*, 233 (1964) and *10*, 141 (1967)
100. Harari, H.: Phys. Lett. *86B*, 83 (1979)
101. Shupe, M. A.: Phys. Lett. *86B*, 87 (1979)
102. Barut, A. O.: Phys. Lett. *73B*, 310 (1978)
103. Barut, A. O.: Phys. Rev. Lett. *42*, 1251 (1979)
104. Steigman, G., Olive, K., Schramm, D. N.: Phys. Rev. Lett. *43*, 239 (1979)
105. Schramm, D., Steigman, G.: Phys. Lett. *87B*, 141 (1979)
106. Kolb, E. W., Goldman, T.: Phys. Rev. Lett. *43*, 897 (1979)
107. Georgi, H., Glashow, S. L.: Phys. Rev. Lett. *32*, 438 (1974)
108. Georgi, H., Quinn, H. R., Weinberg, S.: Phys. Rev. Lett. *33*, 451 (1974)
109. Gross, D. J., Wilczek, F.: Phys. Rev. Lett. *30*, 1343 (1973)
110. Politzer, H. D.: Phys. Rev. Lett. *30*, 1346 (1973)
111. Frampton, P. H., Nandi, S., Scanio, J. J. G.: Phys. Lett. *85B*, 225 (1979)
112. Jørgensen, C. K.: Naturwissenschaften *67*, 35 (1980)
113. Franklin, J.: Phys. Rev. *D20*, 1742 (1979)
114. Glück, M.: Phys. Lett. *87B*, 247 (1979)
115. Lipkin, H. J.: Phys. Lett. *89B*, 358 (1980)
116. Taylor, J. G.: Phys. Lett. *88B* 291 (1979)
117. Quigg, C., Thacker, H. B., Rosner, J. L.: Phys. Rev. *D21*, 234 (1980)
118. Casalbuoni, R., Gatto, R.: Phys. Lett. *88B*, 306 (1979)
119. Krolikowski, W.: Phys. Lett. *90B*, 241 (1980)
120. Elbaz, E., Meyer, J.: preprints from Université Lyon 1 (Villeurbanne)
121. Ballentine, L. E.: Rev. Mod. Phys. *42*, 358 (1970)
122. Jørgensen, C. K.: Theoret. Chim. Acta *34*, 189 (1974)
123. Fechner, W. B., Joss, P. C.: Nature *274*, 347 (1978)
124. Marinescu, N., Stech, B.: Naturwissenschaften *63*, 155 (1976)
125. Hurst, G. S., et al.: Rev. Mod. Phys. *51*, 767 (1979)

Gas Phase Photoelectron Spectra
of *d*- and *f*-Block Organometallic Compounds

Jennifer C. Green

Inorganic Chemistry Laboratory, South Parks Road, Oxford, OX1 3QR, G.B.

Organometallic compounds display a wide variety of non-classical bonding interactions largely as a result of the ability of the π-electrons of organic groups to form covalent bonds with d-block transition metals. Photoelectron spectroscopy has proved an invaluable tool in confirming and extending ideas on the nature of these interactions. Studies reported are all on gas phase species for which detailed information on the valence electrons may be obtained. Relative intensity changes observed on varying the ionizing radiation between He(I) and He(II) are of considerable assistance in assigning photoelectron bands and in providing information on the localization of the molecular orbitals from which the associated electrons are ionized. Most assignments are discussed in relation to simple, qualitative molecular orbital schemes. Comparisons are made between bonding of different metals and different organic groups. Particular emphasis is placed, where information is available, on the differences between d- and f-block metals in this class of compounds. The detailed understanding which has resulted from a combination of photoelectron spectroscopy and theoretical studies in this area has led to reliable models which should prove useful to the experimental chemist.

Table of Contents

List of Abbreviations

acac	2,4-pentanedionate	L.U.M.O.	lowest unoccupied molecular orbital
accp	acetylcyclopentadienyl		
An	actinide	Me	methyl-CH_3
anis	anisole	mecp	methylcyclopentadienyl
A.O.	atomic orbital	mes	mesitylene
bz	benzene	M.O.	molecular orbital
cht	cycloheptatrienyl	P.E.	Photoelectron
C.I.	configuration interaction	pmcp	pentamethylcyclopentadienyl
Clcp	chlorocyclopentadienyl	S.C.F.	self-consistent field
cot	cyclooctatetraene	S-O	spin-orbit
cp	cyclopentadienyl	tfa	1,1,1-trifluoro-2,4-pentadionate
hfa	1,1,1,5,5,5-hexafluoro-2,4-pentanedionate	THF	tetrahydrofuran
		tmecp	tetramethylethylcyclopentadienyl
hmbz	hexamethylbenzene	tmh	2,2,6,6-tetramethyl-3,5-heptanedionate
I.E.	ionization energy		
Ln	Lanthanide	tol	toluene

A. Introduction

Photoelectron (P.E.) spectroscopy is now a well established technique for obtaining information on chemical bonding both in gas phase molecules and in the solid state. This work reviews, in some detail, P.E. studies on organic compounds of the d- and f-block transition metals. The volatility and the extensive and subtle variety of these compounds have made gas phase P.E. spectroscopic investigation of their bonding especially fruitful. That there now exists a coherent picture of the binding of transition metals to organic groups is in part due to the symbiotic relationship between P.E. spectroscopy and bonding theory.

Two excellent reviews[1,2] on P.E. spectroscopy of transition metal compounds have appeared recently, and more general aspects of the technique are continuously reviewed[3-6]. Detailed exposition of fundamental principles of P.E. spectroscopy and its scope may be found in these and in the several books available on the subject[7-12]; this introduction will be restricted to particular general points of special relevance to the area covered.

I. The Photoelectric Experiment

In a photoelectric experiment monochromatic radiation, $h\nu$, causes ionization of matter, and the properties of the ejected electrons are measured. Radiation is of three main types; X-ray, U.V. (normally from an inert gas discharge lamp), and synchrotron radiation. The matter is usually in the solid or gaseous state, though some experiments have also been carried out on liquids and on matrix isolated species. Measurement of the kinetic energy, $\frac{1}{2}mv^2$, of the ionized electrons and use of the Einstein equation

$$h\nu = I.E. + \frac{1}{2}mv^2$$

gives access to the ionization energies, I.E., of molecules, or binding energies of solids. The electron flux intensity and its angular variation is also of interest.

This review is concerned with P.E. spectroscopy of gas phase molecules ionized by He(I) (21.22 eV) and He(II) (40.81 eV) radiation. The vast majority of observations are made at right angles to the radiation where the electron flux is maximum.

II. Interpretation of a P.E. Spectrum

A photoelectron spectrum consists of a plot of the number of electrons with energy $\frac{1}{2}mv^2$, versus that energy. The electron kinetic energies may be converted into ionization energies (I.E.) of the molecule, M, which is a measure of the energy of the process

$$M \xrightarrow{h\nu} M^+.$$

As the molecular ion, M^+, may be formed in excited states as well as in the ground state, a series of bands is obtained for the P.E. spectrum.

For small molecules, the bands show vibrational fine structures, but for the organometallics dealt with in this review, vibrational structure is rarely resolved. It does however affect the band width: when the molecular geometry is largely unchanged on ionization (e.g. on ionizing an electron from a non-bonding molecular orbital, M.O.), the band will be sharp as the probability of the ion being in excited vibrational states is low; however, if the molecular geometry changes on ionization (as when a bonding or anti-bonding electron is ionized), broad bands are obtained as the molecular ion is formed in a multiplicity of excited states.

The primary information obtained from P.E. spectroscopy is the energy of the molecular ion states relative to the ground state molecule and, of course, relative to each other. If P.E. data is to be used to check theoretical calculations on the electronic structure of molecules, the calculation must engender predictions of ionization energies. There are three methods used for calculating ionization energies which are described below.

III. Some Methods of Calculating Ionization Energies

1. Direct Method

In this procedure Hartree-Fock self-consistent field (S.C.F.) or other calculations are carried out on the molecular ground state and on the relevant ionic states, and the calculated energy differences are compared with experimental ionization energies. In the case of S.C.F. calculations this is termed the ΔSCF method.

2. Koopmans' Theorem

Koopmans' theorem[13] states that the ionization energies of a molecule are equal to the negative of the S.C.F. orbital energies, ϵ_i that is

I.E. $= - \epsilon_i$.

This theorem is recognized as an approximation as, apart from the inaccuracies inherent in the S.C.F. method (such as neglect of electron correlation and relativistic effects), it assumes that the molecular orbitals are the same for the molecule and the molecular ion. Many ΔSCF calculations have shown (see for example [14–16]) that if an electron is removed from a metal localized orbital, considerable charge migration towards the metal occurs; this is termed relaxation. These relaxation effects give ionization energies smaller values than those expected on the basis of Koopmans' theorem. For ionization from ligand based orbitals, relaxation effects are smaller and more constant.

If these deficiencies in Koopmans' theorem are recognized, it is possible to argue that relaxation effects are relatively constant for related orbitals in chemically similar molecules (see for example [17-19]), which enables empirical corrections to be made to Koopmans' theorem.

Use of Koopmans' theorem in conjunction with parameterized, non-S.C.F. calculations is often successful at predicting correct ionization energy sequences.

3. Transition State Method

This method may be used in conjunction with the various types of Xα calculation. This enables the difference between two state energies to be estimated by calculation of a transition state, which involves occupation numbers half-way between the initial and final states. In the case of ionization energies, this corresponds to removal of half an electron from the appropriate orbital.

It is outside the scope of this review to attempt to evaluate the relative merits of the various types of molecular calculation and their ability to predict I.E. accurately. It is instructive, however, to compare some predictions for ferrocene, which has been the subject of many a theoretical investigation. The results obtained are presented in Table 1. It is clear from this survey that, for molecules of this size, assignment of P.E. spectra cannot be made on the basis of calculation alone: theoretical estimates give a good guide to ordering of ionization energies, but empirical criteria are necessary for certain assignment.

Table 1. Comparison of theoretical estimates of ionization energies (eV) of ferrocene

M.O. Ion states	Koopmans' theorem[a] and direct methods				Transition state		Experimental
	ΔIEHT[b 44)]		ΔSCF [15)]	ΔSCF [36)]	Xα-SW[c 43)]	Xα (HFS)[d 37)]	
$e_{2g}\,^2E_{2g}$	x	(11.92)	8.3 (14.4)	5.7 (11.8)	8.5	6.7	6.8
$a_{1g}\,^2A_{1g}$	x + 0.4	(11.6)	10.1 (16.6)	7.5 (14.3)	7.9	6.7	7.2
$e_{1u}\,^2E_{1u}$	x + 1.9	(12.16)	11.1 (11.7)	8.9 (9.5)	9.3	8.1	8.8
$e_{1g}\,^2E_{1g}$	x + 2.5	(12.48)	11.2 (11.9)	8.8 (9.5)	9.7	8.6	9.3

[a] Koopmans' theorem values are given in parentheses
[b] Iterative Extended Huckel Theory (IEHT)
[c] Xα-Scattered Wave (Xα-SW) calculation using a muffin-tin approximation
[d] Hartree-Fock-Slater (HFS) calculation using Xα exchange but no muffin-tin restrictions

IV. Band Intensities

For detailed discussion of theoretical treatments of photoionization cross sections, the reader is referred to more general accounts [6,11]. Here we are concerned with the band intensities observed in a conventional P.E. experiment and their variation with photon energy, in so far as this data can aid assignment of P.E. spectra and give information on the nature of the ground state wave-function.

Experimentally, various trends in band intensities are observed:

i) for photoionization of metal nd orbitals there is a significant increase in relative band intensity as the principal quantum number, n, increases. This is known as "the heavy atom effect".

ii) for ionization of M.O. with appreciable metal d character, band intensities often increase markedly compared with ligand based M.O. bands when the ionizing radiation is changed from He(I) to He(II)[2]. The increase is significant with respect to hydrocarbon valence orbitals[20] and most pronounced when the metal is attached to a ligand with principal quantum number greater than 2 such as sulphur, phosphorus or the halogens[21]. There is very little change in band intensity with respect to oxygen ligand bands, and the metal d-bands decrease in intensity compared with fluorine bands[22,23].

iii) The bands associated with ionization of f-electrons are very low in intensity in He-I spectra[24] but increase significantly in relative intensity in He-II spectra.

Empirical observations such as these are best rationalized using what has become known as "the Gelius Model"[25]. In this the one electron ionization cross section σ_j for the jth M.O. is given by

$$\sigma_j = \sum_{A,i} P_{j,i_A}\, \sigma_i^A$$

where the summation extends over atomic orbitals, ϕ_i^A, on different atomic centres, A, which contribute to the M.O., ϕ_j. The σ_i^A are one electron atomic cross sections and the P_{j,i_A} are factors describing effective occupancy of the M.O. They are best equated to the square of the atomic orbital, A.O., contribution, c_{j,i_A}^2, to M.O. ϕ_j where

$$|\phi_j\rangle = \sum_{i_A} c_{j,i_A}\, |\phi_i^A\rangle$$

which implies that the overlap population does not contribute to the ionization cross section[6]. In this model the molecular ionization cross sections are directly related to the atomic cross sections of the A.O. contributing to the appropriate M.O. Though many attempts have been made to calculate the atomic cross sections, σ_i^A,[6] for our purposes they are best regarded as empirical parameters and their variation with photon energy determined experimentally.

In order to relate photoionization cross sections to band intensities observed at 90° to the photon flux, account must be taken of the fact that the electron flux is anisotropic. The angular variation for unpolarized radiation is given by

$$\frac{d\sigma_j^v}{d\Omega} = \frac{\sigma_j^v}{4\pi}\left[1 - \frac{\beta_j^v}{4}(3\cos^2\theta - 1)\right]$$

where θ is the angle between the directions of photon and electron propagation, and β_j^v is an energy dependent asymmetry parameter. If $\theta = 90°$, the intensity is clearly dependent on β_j^v, and if variations in band intensity are discussed in terms of σ_j^v alone, variations in β_j^v are being neglected.

Though the model clearly has quantitative potential, its main application in transition metal P.E. spectroscopy has been qualitative in identifying the predominant A.O. character in the M.O. giving rise to particular P.E. bands. It is noteworthy that information is given on the nature of the vacated M.O.; therefore band intensities are determined primarily by the ground state molecular wavefunction, whereas the energy information of a P.E. experiment relates principally to the excited states.

The Gelius model is at the end of a long chain of successive approximations in the theory of photoionization cross sections, and its range of applicability is uncertain. It is understood to work best at high photon (and electron) energies, and its extension to spectra produced by He(I) and He(II) radiation is problematic. Its justification in these circumstances is largely empirical and qualitative; its predictions as to the localization of M.O. normally conform with theoretical expectations. One spectacular example of its breakdown, however, concerns the P.E. spectrum of O_2. Under He(II) excitation the $(\pi_g)^{-1}/(\pi_u)^{-1}$ intensity ratio is three times higher than expected on a purely statistical (electron occupancy) basis[26], but π_g and π_u have identical atomic constitution, to a first approximation.

It should also be recognized that autoionization can severely modify band intensities and may occur at low photon energies.

It is hoped that further studies using a wider variety of photon energies, as provided by synchroton radiation, may help identify the degree to which inferences from intensity changes are valid.

V. Photoionization of Open Shell Systems

For open shell molecules, the correlation of one P.E. band per occupied M.O. breaks down as ionization of the open shell itself, or of subsequent closed shells, may give rise to a multiplicity of ion states. The rules for predicting the ion states formed and the relative intensities of the associated P.E. bands have been derived[27] and are summarized below.

i) If a closed shell is ionized, all states arising from the coupling of the positive hole with the open shell state will be realized, the relative cross sections for the production of these states being in proportion to the total (i.e. spin-orbital) degeneracies.

ii) If orbitals belonging to different sub-shells are assumed to have the same one-electron cross sections, the integrated ionization cross section of a particular sub-shell is simply proportional to the occupancy of that orbital in the sub-shell of the molecule.

iii) If an open shell is ionized, the relative probabilities of producing different ionic states will reflect the fractional parentage coefficients which may, but will not in general, be proportional to spin-orbital degeneracies.

iv) If a molecule contains two or more open-shells, it is necessary to consider the coupling which already exists between the different open-shells in the molecule. The probability of ionization is usually expressed in terms of Racah coefficients.

Examples of applications of these rules are given in Table 3.

VI. Presentation of Results

In the subsequent sections, the P.E. spectra of d- and f-block organometallics and their likely assignments are presented. P.E. spectra of organometallics show an ubiquitous band, normally lying between 11 and 16 eV, which is due to ionization from the σ-structure of the organic ligand. Though these bands exhibit features characteristic of the organic groups from which they arise, very few deductions have been made from them. In the interests of brevity, these bands are frequently omitted from the figures, and also the tables, and attention in the discussion is focussed on the higher lying valence levels of these molecules, which dominate the metal-ligand bonding and the reactivity of these compounds. More detailed information on these bands may be obtained from the original literature.

Some of the figures show level diagrams for key molecules. The purpose of these diagrams is to show symmetry correlations and metal-ligand interactions. The levels do not represent S.C.F. orbital energies, rather they indicate the ionization energies of orbitals; they only relate to orbital energies within the limitations of Koopmans' theorem. They are therefore referred to as interaction diagrams, rather than M.O. diagrams, in order to remove any possible ambiguity.

In the context of these correlations, a M.O. in an organometallic will often be referred to by the nomenclature of its principal atomic or ligand orbital constituent. This convenient christening of M.O. is in no way intended to minimize the covalent interaction between metal and ligand, but it is adopted to emphasize correlations between M.O. in molecules of different symmetry.

B. Sandwich Compounds

To a large extent, metal sandwich compounds, which are those where a metal atom lies between two parallel carbocyclic rings, have a common orbital structure. This is a consequence of the similarity of the lower π-orbitals of the rings themselves; it is effectively independent of the actual molecular symmetry. In labelling the molecular orbitals and ion states for this class of molecule, we will assume an infinite axis of rotation[28,29] and, for the bis-cyclopentadienyl and bis-arene compounds, a centre of inversion. Interaction diagrams and sketches of the M.O. are given in Fig. 1 as an aid to keeping track of the metal ligand interactions possible for these molecules. As explained in section A. VI, these do not correspond to the results of the more accurate calculations rather they may be regarded as ionization energy diagrams, reflecting the relative ease of ionizing electrons from the various orbitals.

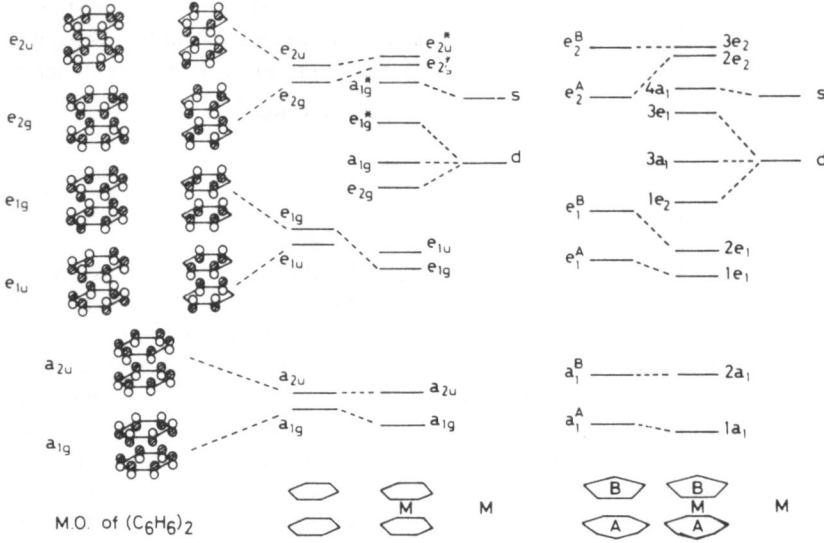

M.O. of $(C_6H_6)_2$

Fig. 1a, b. Interaction diagrams for sandwich molecules: (a) a bis-η-arene metal compound, (b) a mixed ring sandwich compound, $M(\eta\text{-}C_5H_5)(\eta\text{-}C_7H_7)$

Assignments of the p.e. spectra of the sandwich compounds are discussed below according to their electron number.

I. 18 Electron Compounds

1. Bis-Arene Compounds

Although ferrocene is historically the parent of this class of compounds, in many ways the P.E. spectra of the bis-η-arene metal compounds are easier to interpret. Representative examples of the chromium, molybdenum and tungsten derivatives are given in Fig. 2[17,20,28,30], and the lower ionization energy data is tabulated (Table 2). The upper regions of the spectra (I.E. > ca. 10 eV) closely resemble those of the free arenes; the ionization energies and band contours are similar, but the bands are broader in the complexes than in the free ligands. These bands are assigned in a similar manner of those of the free arenes.

The lowest I.E. band in all cases is assigned to the $^2A_{1g}$ ion state. This is consistent with e.s.r. evidence on $[Cr(\eta\text{-}C_6H_6)_2]^+$ [31–33] and other bis-η-arene complexes[34]. This band is extremely sharp for an organometallic compound indicating ionization from a non-bonding orbital; calculations consistently predict the a_{1g} orbital of metal sandwich compounds to be practically pure metal d_{z^2} in character as a consequence of

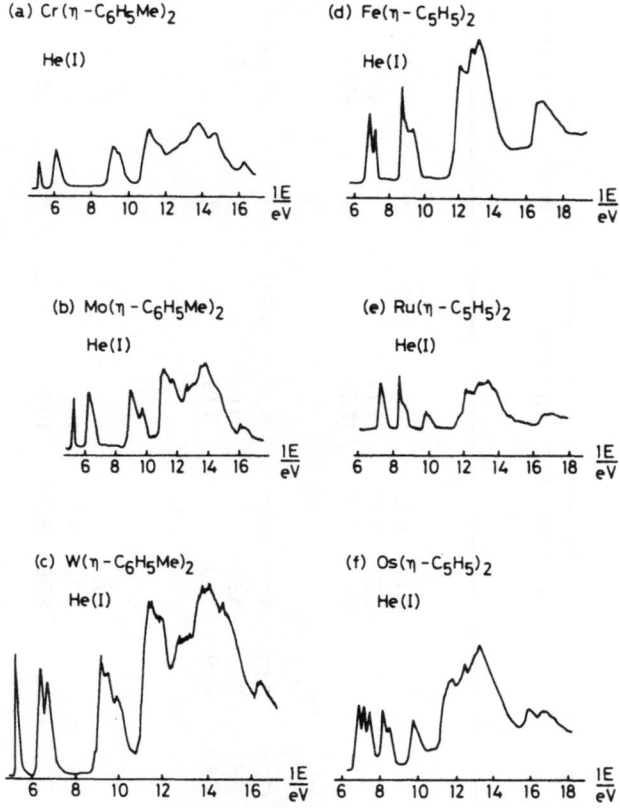

Fig. 2a–f. He(I) P.E. spectra of some 18 electron compounds: **a** Cr(η-C$_6$H$_5$Me)$_2$; **b** Mo(η-C$_6$H$_5$Me)$_2$; **c** W(η-C$_6$H$_5$Me)$_2$; **d** Fe(η-C$_5$H$_5$)$_2$; **e** Ru(η-C$_5$H$_5$)$_2$; **f** Os(η-C$_5$H$_4$Me)$_2$

the near zero overlap for this metal orbital with ring $a_1(\pi)$ orbitals (see for example[15,35–41]).

The second band for the chromium and molybdenum derivatives is assigned to the $^2E_{2g}$ ion state. In the case of tungsten, this band is split into two spin-orbit components, $^2E_2$ ($\frac{5}{2}$) and $^2E_2$ ($\frac{3}{2}$), as a result of the greater spin-orbit coupling constant for the third-row transition metal.

The subsequent band, lying between 8 and 10.5 eV, is complex and is assigned to the $^2E_{1u}$ and $^2E_{1g}$ ion states. The splitting between these states is barely resolved in the chromium case but increases for molybdenum and tungsten. The lower I.E. band has a sharp leading edge whereas the higher I.E. band has a profile more characteristic of a bonding orbital. On these grounds the higher ionization is assigned to the $^2E_{1g}$ ion state, as the e_{1g} orbital is assumed to be more involved in metal-ligand bonding than the e_{1u} orbital on account of the metal d-character of the former orbital; the lower of the two bands is assigned to the $^2E_{1u}$ ion state[17].

Table 2. Ionization energy data for transition metal sandwich compounds

Compound	Electron number	Molecular ground state	Ionization energies[a] (eV)					e_{1u} or e_{1b}	e_{1g} or e_{1a}	He(II)	References
V(cp)₂	15	$^4A_{1g}$				$^3A_{1g}$ 6.78	$^3E_{2g}$ 6.78	8.40 8.65	8.79 9.02	II	48, 49
V(mecp)₂	15	$^4A_{1g}$				6.60	6.60	(8.03) 8.39	8.73		49
V(pmcp)₂	15	$^4A_{1g}$				5.87	5.87	(7.08) 7.27	(7.69)	II	48
Cr(cp)₂	16	$^3E_{2g}$	$^4A_{2g}$ 5.71	$^2E_{1g}$ 7.04	$^2E_{2g}$ 7.30	$^2A_{1g}$ (7.58)	$^2A_{2g}$ (7.58)	8.55 (8.87)	(9.25)	II	47–49, 57
Cr(mecp)₂	16	$^3E_{2g}$	5.53	6.85	7.10	7.41	7.41	8.32	9.02		49, 57
Cr(pmcp)₂	16	$^3E_{2g}$	4.93	6.18	6.34	6.69	6.69	7.27	7.89	II	48
Ti(bz)₂	16	$^1A_{1g}$					$^2E_2$ 5.5–6.0	9.0	9.3		59
Ti(tol)₂	16	$^1A_{1g}$					5.4		10.2		59
Ti(cp)(cht)	16	$^1A_{1g}$					6.83	8.71 9.1	10.2		17
Zr(cp)(cht)	16	$^1A_{1g}$					6.94	8.78 9.23	10.3		52
V(mes)₂	17	$^2A_{1g}$			$^1A_1$ 5.61	$^3E_2$ 5.33	$^1E_2$ 6.08	8.75[b]			17
Nb(bz)₂	17	$^2A_{1g}$			5.57	6.17	6.67	—[c]		II	53
Nb(tol)₂	17	$^2A_{1g}$			5.43	5.90	6.46	9.14	9.46	II	53
Nb(mes)₂	17	$^2A_{1g}$			5.18	5.59	6.15	8.59	9.19	II	53
Ti(cp)(cot)	17	$^2A_{1g}$			5.67	7.62	7.62	8.63 8.93	10.51		17
V(cp)(cht)	17	$^2A_{1g}$			6.42	6.77	7.28	8.66 8.99	10.2		17
Nb(cp)(cht)	17	$^2A_{1g}$			5.98	7.11	7.50	8.78 9.13	10.4		52
Cr(cp)(bz)	17	$^2A_{1g}$			7.15	6.20	7.15	8.76 9.17	9.68		17, 28
Mo(cp)(bz)	17	$^2A_{1g}$			6.46	6.46	7.25	8.66	9.89		42

Compound	n	Ground state	$^5E_{1g}$ / $^3E_{2g}$ / $^2A_1$	$^5A_{1g}$ / $^3A_{2g}$ / $^2E_2$	$^5E_{2g}$ / $^1E_{1g}$	$^1A_{1g}$	$^1E_{2g}$					pol	Ref
Mn(cp)$_2$[d]	17	$^6A_{1g}$	6.91	10.10	10.51			8.76[b]				II	47–49
Mn(mecp)$_2$[d]	17	$^6A_{1g}$	6.58	9.90	10.23			8.42[b]				II	48, 49
Mn(pmcp)	17	$^2E_{2g}$	5.33	5.72	6.37	6.50	6.72	7.26		7.95		II	48
Mn(mecp)$_2$[d]	17	$^2E_{2g}$	6.01	—	7.15	—	(7.36)				(8.16)	II	48, 49
Mn(cp)$_2$[d]	17	$^2E_{2g}$	6.26									II	48, 49
			$^2A_1$	$^2E_2$									
Fe(cp)$_2$	18	$^1A_{1g}$	7.23	6.88				8.72	(8.87)	9.14	9.39	II	7, 45–48
Fe(mecp)$_2$	18	$^1A_{1g}$	7.06	6.72				8.53	(8.73)	9.17			46
Fe(pmcp)$_2$	18	$^1A_{1g}$	6.28	5.88				7.31		8.08		II	48
Fe(tmecp)$_2$	18	$^1A_{1g}$	6.31	6.00				7.38	(7.52)	8.15			63
Fe(Clcp)$_2$	18	$^1A_{1g}$	7.37	7.06				8.71	9.09	9.49			46
Fe(accp)$_2$	18	$^1A_{1g}$	7.36	7.06				8.97		9.61			64
Ru(cp)$_2$[e]	18	$^1A_{1g}$	7.45	7.45	(7.68)			8.51	(8.80)	9.93	(10.23)	II	46, 48
Ru(mecp)$_2$[e]	18	$^1A_{1g}$	7.25	7.25				8.24	(8.40)	9.76			46
Os(mecp)$_2$	18	$^1A_{1g}$	7.21	6.93	7.55[f]			8.26	8.68	9.90			46
Cr(bz)$_2$	18	$^1A_{1g}$	5.45	6.46				9.56	9.80				20, 28
Cr(tol)$_2$	18	$^1A_{1g}$	5.24	6.19				9.16		9.53		II	28
Cr(mes)$_2$	18	$^1A_{1g}$	5.01	5.88				8.90[b]					17
Mo(bz)$_2$	18	$^1A_{1g}$	5.52	6.59				9.47	10.15				17
Mo(tol)$_2$	18	$^1A_{1g}$	5.32	6.33				9.05		9.75			17
Mo(mes)$_2$	18	$^1A_{1g}$	5.13	6.03				8.63		9.31			17
W(bz)$_2$	18	$^1A_{1g}$	5.40	6.56	6.99[f]			9.58		10.3		II	30, 42
W(tol)$_2$	18	$^1A_{1g}$	5.25	6.28	6.65[f]			9.32	(9.59)	10.08		II	42
W(mes)$_2$	18	$^1A_{1g}$	5.18	6.16	6.51[f]			8.77	(9.06)	9.48		II	30, 42
W(anis)$_2$	18	$^1A_{1g}$	5.49	6.45	6.74								42
Cr(cp)(cht)	18	$^1A_{1g}$	5.59	7.19				8.69	(9.00)	10.4			17
Mo(cp)(cht)	18	$^1A_{1g}$	5.87	7.55				8.93	(9.28)	10.4			52
Mn(cp)(bz)	18	$^1A_{1g}$	6.36	6.72				8.75	(9.25)	9.79			28

Table 2 (continued)

Compound	Electron number	Molecular ground state	Ionization energies[a] (eV)		e_{1u} or e_{1b}	e_{1g} or e_{1a}	He(II)	References
			$^1A_1$	other d-bands				
Co(cp)$_2$	19	$^2E_{1g}$	5.56	7.18, 7.63, 8.01, 9.88	8.66 (8.94)	(9.31)	II	48, 49
Co(mecp)$_2$	19	$^2E_{1g}$	5.37	6.97, 7.43, 7.80, 9.59	8.40 (8.64)	(9.03)		49
Co(pmcp)$_2$	19	$^2E_{1g}$	4.71	6.39	7.55	8.30	II	48
Fe(cp)(hmbz)	19	$^2E_{1g}$	4.93	6.09, 7.15, 7.42	8.14 (8.43)	9.03	II	60
			$^2E_{1g}$	other d-bands				
Ni(cp)$_2$	20	$^3A_{1g}$	6.51	8.43, 9.22, 10.33	8.78[b]		II	48, 49
Ni(mecp)$_2$	20	$^3A_{1g}$	6.36	(8.30) 9.05, 10.00	8.53[b]			49
Ni(pmcp)$_2$	20	$^3A_{1g}$	5.82	7.47, 8.40	7.71[b]		II	48

[a] Where possible the d-ionization bands are classified according to the ion state whereas the ligand π-ionization bands are classified according to the M.O. from which the electron is removed

[b] different e_1 ionizations may not be distinguished

[c] e_1 ionizations obscured by P.E. spectrum of free ligand

[d] high spin-low spin gas phase equilibrium

[e] $^2A_{1g}$ and $^2E_{2g}$ ion states overlap

[f] spin-orbit splitting of $^2E_2$ state

Further confirmation of these assignments comes from the examination of the band ionization cross sections [28] and their variation with photon energy [20]. Lloyd et al.[20] point to the close correlation between the intensity changes for these first four bands and the proposed orbital character. The changes are interpreted as a reduction of intensity of ligand levels or as an increase in intensity for metal bands, or both, on changing from He(I) to He(II) radiation. For $Cr(\eta\text{-}C_6H_6)_2$ Lloyd found that, if the intensities are referred to that of band 1 as a standard, then band 2 decreased in intensity and bands 3 and 4 decreased much more. This is in agreement with the hypothesis that the first band is due to·ionization from a_{1g}, which is almost pure metal, the second due to ionization from e_{2g}, which is of mixed metal-ligand character. This pattern of intensity changes is also shown for other bis-arene compounds[42]. Where the bands assigned to the $^2E_{1u}$ and $^2E_{1g}$ ion states are clearly separated, the higher I.E. band shows an intensity increase relative to the lower band consistent with metal d-character in the orbital from which the higher band arises, which confirms its assignment to the $^2E_{1g}$ ion state.

Whereas the intensity of the first two bands in the He(II) spectra reflect their orbital degeneracies[20], the relative cross-sections in He(I) spectra ($a_{1g} : e_{2g} \approx 1 : 4$) are consistent with the above interpretation if ligand orbitals have larger cross sections than metal d-orbitals and if the cross sections of the M.O. are related to the cross sections of their constitutent A.O.[28].

2. Bis-Cyclopentadienyl Compounds

The d^6 bis-cyclopentadienyl complexes constitute those of the iron group and include ferrocene. The reported P.E. spectra of these complexes, which include substituted derivatives, are tabulated (Table 2); representative spectra of the low I.E. region of the spectra of ferrocene, ruthenocene and 1,1-dimethylosmocene are shown in Fig. 2. The spectrum of ferrocene has been recorded by many authors[7,45−48], and the experimental results are in good agreement. Rabalais[47] claims to resolve vibrational fine structure of spacing 35 meV on the first three bands, but Evans et al.[49] failed to reproduce this even with a working resolution on CH_3I of 19 meV. The existence of these vibration progressions must therefore remain in doubt.

There is a general consensus on the assignment of the first two bands of ferrocene to the $^2E_{2g}$ (6.88 eV) and $^2A_{1g}$ (7.23 eV) ion states. The ordering of these states is reversed compared with the bis-arene cations. This assignment is consistent with the $^2E_{2g}$ ground state for $[Fe(\eta\text{-}C_5H_5)_2]^+$ suggested by esr[50,51] and gives the most reasonable explanation of the relative intensity of the first two bands ($\sim 2.5 : 1$)[46]. It is further confirmed by He(II) studies on ferrocene[48] where the intensity of the second band increases relative to the first and the intensity ratio approaches $2 : 1$, this is consistent with the second band being due to ionization from the a_{1g} orbital.

The third and fourth bands are assigned by Evans et al.[46] to the ion states $^2E_{1u} < {}^2E_{1g}$, whereas Rabalais[47], on the grounds of the vibrational progression, proposed the reverse ordering. He(II) studies confirm the former suggestion as the fourth band increases in intensity relative to the third at the higher photon energy[48].

In the P.E. spectrum of ruthenocene [46] the $^2E_{2g}$ and $^2A_{1g}$ ion states are unresolved, and the separation of the $^2E_{1u}$ and $^2E_{1g}$ states, which are assigned in the same order as for the ferricenium cation, increases. Comparable intensity changes have been recorded in the He(II) spectrum [48].

The spectrum of $Os(\eta\text{-}C_5H_4Me)_2$ [46] shows three bands in the low I.E. region. From the magnitude of the spin-orbit coupling, Evans et al. argue that the preferred assignment is $^2E\left(\frac{5}{2}\right) < {}^2A_1 < {}^2E\left(\frac{3}{2}\right)$. The change in the $^2E_{2g} - {}^2A_{1g}$ energy separation on passing from Fe to Ru to Os is attributed to an $a_1(d) - e_2(d)$ orbital energy difference and is explained by stabilization of the e_2 orbital due to increased covalence going down the group. The energy of the $^2E_{1u}$ state varies very little whereas the $^2E_{1g}$ ion state is progressively stabilized on passing from Fe to Ru to Os. This trend is attributed to an increased stabilization of the e_{1g} M.O. through $d - \pi$ mixing and further suggests greater covalency in the heavier congeners.

The combined cross section for the $^2E_{2g}$ and $^2A_{1g}$ bands steadily increases relative to the cross section of the ligand $^2E_{1u}$ state on going from iron through to osmium. This is attributed in part to an increase with principal quantum number of the intrinsic metal d cross section, the so-called heavy atom effect (see Sect. A.IV).

The high I.E. bands all appear to be characteristic of the ligand species. The main band system occurs in the region 11.5–15 eV in the cyclopentadienyl compounds and between 11 and 15.5 eV in the methylcyclopentadienyls, the low energy shoulder being due to ionization of electrons localized on the CH_3 group.

3. Mixed Sandwich Compounds

The d^6 mixed sandwich compounds which have been reported are $Mn(\eta\text{-}C_5H_5)$-$(\eta\text{-}C_6H_6)$ and $M(\eta\text{-}C_5H_5)(\eta\text{-}C_7H_7)$ where $M = Cr$ and Mo [17,133]. The low energy bands may be assigned as for the arene sandwich compounds, which leads to an ion state ordering $^2A_1 < {}^2E_2 < {}^2E_1 < {}^2E_1$. The first ionization band is sharp, indicating the non-bonding nature of the $a_1(d_{z^2})$ orbital in these compounds. The assumed trends in orbital energies are discussed below in Sect. B. VII.

II. 17 Electron Compounds

A variety of ground states are shown for sandwich compounds with a d^5 configuration. Bis-arene metal compounds have $^2A_{1g}$ ground states corresponding to an $e_{2g}^4 a_{1g}^1$ configuration [33,53]. A similar configuration for the mixed sandwich compounds $M(\eta\text{-}C_5H_5)(\eta\text{-}C_6H_6)$ where $M = Cr$ and Mo, $M(\eta\text{-}C_5H_5)(\eta\text{-}C_7H_7)$ where $M = V$ and Nb and $Ti(\eta\text{-}C_5H_5)(\eta\text{-}C_8H_8)$ [34,54] confers a $^2A_1$ ground state. Manganocene, however, has a $^6A_{1g}$ ground state arising from the high spin configuration $e_{2g}^2 a_{1g}^1 e_{1g}^2$, whereas decamethylmanganocene is low spin with a $^2E_{2g}$ ground state $(e_{2g}^3 a_{1g}^2)$. 1,1-dimethyl-manganocene exists as a high spin-low spin mixture at ambient temperatures [49,55].

Ion states that are accessible by one electron processes from these various ground states are given in Table 3, together with their predicted relative intensities. These predictions are made using Cox's rules discussed earlier in Sect. A.V. Also tabulated are the ligand field energies. A more extensive tabulation is given by Warren [29].

1. $^2A_1$ Ground States

Figure 3 gives a selection of spectra obtained for molecules of this class. Ionization of an $e_2^4 a_1^1$ configuration is expected to give rise to three states $^1A_1(e_2^4)$ $^3E_2$ and $^1E_2(e_2^3 a_1^1)$. The $^3E_2$ state will lie lower in energy than the $^1E_2$ state.

The spectra of M(arene)$_2$ and M(η-C$_5$H$_5$) (η-C$_7$H$_7$) where M = V and Nb show three clearly discernable bands in the low energy region [17,52]. They may be distinguished by their profiles, even when overlapping; the $^2A_1$ ionization is sharp and of lowest intensity while the E$_2$ ionizations are broader, the $^3E_2$ state having a more intense band than the $^1E_2$ state. The cation $[V(C_6H_3Me_3)_2]^+$ shows a $^3E_2$ ground state whereas the other ions have a $^1A_1$ ground state.

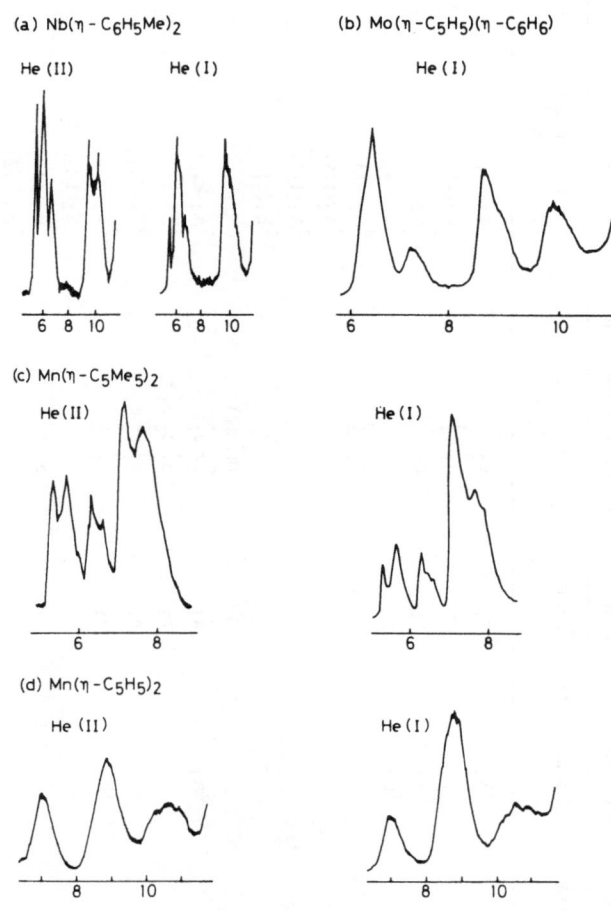

Fig. 3a–d. P.E. spectra of some 17 electron compounds: **a** Nb(η-C$_6$H$_5$Me)$_2$, He(I) and He(II); **b** Mo(η-C$_5$H$_5$) (η-C$_6$H$_6$) He(I); **c** Mn(η-C$_5$Me$_5$)$_2$, He(I) and He(II); **d** Mn(η-C$_5$H$_5$)$_2$, He(I) and He(II)

Table 3. Ion states, intensities and ligand field energies for ionization of sandwich molecules[a]

Configuration	Ground state	Orbital ionization	Ion configuration	Ion states produced and intensities[b]	First order ligand field energies[c]
$a_1^1 e_2^2$	$^4A_2$	a_1	e_2^2	$^3A_2(k_{a_1})$	$A+4B$
		e_2	$a_1^1 e_2^1$	$^3E_2(2)$	$A-8B+\Delta_2$
e_2^4	$^1A_1$	e_2	e_2^3	$^2E_2$	$3A+12B+4C$
$a_1^1 e_2^3$	$^3E_2$	a_1	e_2^3	$^2E_2(k_{a_1})$	$3A+12B+4C$
		e_2	$a_1^1 e_2^2$	$^4A_2(4/3)$	$3A-12B+\Delta_2$
				$^2A_2(1/6)$	$3A+3C+\Delta_2$
				$^2A_1(1/2)$	$3A-8B+5C+\Delta_2$
				$^2E_1(1)$	$3A-8B+3C+\Delta_2$
				$(^2E_2)$	$3A+12B+4C$
$a_1^1 e_2^2 e_1^{*2}$	$^6A_1$	d	$(a_1^2 e_2^1)$	$^5A_1(k_{a_1})$	$6A-21B+2\Delta_2+2\Delta_1$
		a_1	$e_2^2 e_1^{*2}$	$^5E_2(2)$	$6A-21B+3\Delta_2+2\Delta_1$
		e_2	$a_1^1 e_2^1 e_1^{*2}$	$^5E_1(2k_{e_1}^*)$	$6A-21B+2\Delta_2+\Delta_1$
		e_1	$a_1^1 e_2^2 e_1^{*1}$	$^3E_2(3k_{a_1}/2)$	$6A-8B+5C+\Delta_2$
				$^1E_2(1k_{a_1}/2)$	$6A+7C+\Delta_2$
$a_1^2 e_2^3$	$^2E_2$	e_2	$a_1^2 e_2^2$	$^1E_1(1)$	$6A-16B+7C+2\Delta_2$
				$^1A_1(1/2)$	$6A-16B+9C+2\Delta_2$
				$^3A_2(3/2)$	$6A-16B+5C+2\Delta_2$
$a_1^1 e_2^4$	$^2A_1$	d	(e_2^4)	$(^1A_1)$	$6A+24B+8C$
		a_1	e_2^4	$^1A_1(k_{a_1})$	$6A+24B+8C$
		e_2	$a_1^1 e_2^3$	$^3E_2(3)$	$6A-8B+5C+\Delta_2$
				$^1E_2(1)$	$6A+7C+\Delta_2$
$a_1^2 e_2^4$	$^1A_1$	a_1	$a_1^1 e_2^4$	$^2A_1(2k_{a_1})$	$10A-20B+10C+2\Delta_2$
		e_2	$a_1^2 e_2^3$	$^2E_2(4)$	$10A+10C+\Delta_2$
$a_1^2 e_2^4 e_1^{*1}$	$^1E_1$	a_1	$a_1^1 e_2^4 e_1^{*1}$	$^3E_1(3k_{a_1}/2)$	$15A-13B+12C+2\Delta_2+\Delta_1$
				$^1E_1(k_{a_1}/2)$	$15A-11B+14C+2\Delta_2+\Delta_1$

Ground state: $a_1^2 e_2^4 e_1^{*2}$, $^3A_2$

Orbital	Ion configuration	Ion state	Energy
e_2	$a_1^2 e_2^3 e_1^{*2}$	$^3E_1(3/2)$	$15\,A - 29\,B + 12\,C + 3\,\Delta_2 + \Delta_1$
		$^1E_1(1/2)$	$15\,A - 17\,B + 14\,C + 3\,\Delta_2 + \Delta_1$
		$^3E_2(3/2)$	$15\,A - 29\,B + 12\,C + 3\,\Delta_2 + \Delta_1$
		$^1E_2(1/2)$	$15\,A - 29\,B + 14\,C + 3\,\Delta_2 + \Delta_1$
e_1^*	$a_1^2 e_2^4$	$^1A_1(ke_1^*)$	$15\,A - 20\,B + 15\,C + 2\,\Delta_2$
a_1	$a_1^1 e_2^4 e_1^{*2}$	$^4A_2(4k_{a_1}/3)$	$21\,A - 31\,B + 14\,C + 3\,\Delta_2 + 2\,\Delta_1$
		$^2A_2(2k_{a_1}/3)$	$21\,A - 28\,B + 17\,C + 3\,\Delta_2 + 2\,\Delta_1$
e_2	$a_1^2 e_2^3 e_1^{*2}$	$^4E_2(8/3)$	$21\,A - 43\,B + 14\,C + 4\,\Delta_2 + 2\,\Delta_1$
		$^2E_2^{I}(4/3)$	$21\,A - 34\,B + 17\,C + 4\,\Delta_2 + 2\,\Delta_1$
		$^2E_1(2k_{e_1}^*)$	$21\,A - 31\,B + 18\,C + 3\,\Delta_2 + \Delta_1$
e_1^*	$a_1^2 e_2^4 e_1^{*1}$	$(^2E_2^{II})$	$21\,A - 28\,B + 19\,C + 4\,\Delta_2 + 2\,\Delta_1$
d	$(a_1^2 e_2^3 e_1^{*2})$	$(^2E_2)$	$21\,A - 24\,B + 17\,C + 3\,\Delta_2 + 2\,\Delta_1$
d	$(a_1^1 e_2^4 e_1^{*2})$	$(^2A_2)$	$21\,A - 31\,B + 15\,C + 4\,\Delta_2 + 2\,\Delta_1$
d	$(a_1^2 e_2^3 e_1^{*2})$		

a These expressions are adapted from Warren[29] and Orchard et al.[49]

b Relative transition probabilities according to Cox's rule[27] are given in parentheses k_{a_1} and $k_{e_1}^*$ are the one-electron cross-sections of the a_1(d) and e_1^*(d) orbitals respectively, expressed relative to the e_2(d) cross-section

c A, B and C are the usual Racah parameters of interelectronic repulsion. The orbital parameters Δ_1 and Δ_2 are defined in Sect. B. IX

d These ion states are of low energy but may not be reached by a one electron ionization from the molecular ground state. They may however be involved in configuration interaction with states accessible by such a process

In the case of $M(\eta\text{-}C_5H_5)(\eta\text{-}C_6H_6)$ where $M = Cr$ and Mo [17,28,42], only two bands are resolved; the first band is assigned to the $^3E_2$ state and the second to the $^1E_2$ ion state. The low intensity $^1A_1$ band is assumed to overlap with one of these bands. In the chromium compound it was first assumed on intensity grounds to lie under the $^3E_2$ band [28], but then the assignment was changed [17] to a suggestion of coincidence with the $^1E_2$ ground state as this gave more uniform ionization energy trends (see Sect. B. VII)[1]. If the first band of the molybdenum spectrum is examined carefully, it is seen to consist of two bands, so in this case the $^1A_1$ state overlies the $^3E_2$ state [42].

The P.E. spectrum of $Ti(\eta\text{-}C_5H_5)(\eta\text{-}C_8H_8)$ shows that the ion has a $^1A_1$ ground state, but in this case the bands corresponding to $^3E_2$ and $^1E_2$ are unresolved [17]. Overall the exchange splitting of the E_2 states is found to decrease with increase in ring size. This is attributed to a decrease in the exchange integral $K(a_{1g} \cdot e_{2g})$; the a_{1g} orbital is metal localized whereas, as the ring size increases, the e_{2g} orbital undergoes a decrease in metal content and an increase in ligand π-content [17]. This is discussed further in Sect. B.VII.

2. $^6A_1$ Ground States

The assignment of the P.E. spectrum of manganocene and 1,1'-dimethylmanganocene initially posed problems as the nature of the gas phase ground states was uncertain. Thus Evans et al. [49] assigned $Mn(\eta\text{-}C_5H_5)_2$ on the basis of a $^6A_{1g}$ ground state whereas Rabalais et al. [47] preferred to assume a $^2A_{1g}$ ground state. The spectral pattern differs markedly from those spectra of compounds with well established $^2A_{1g}$ ground states discussed above. Moreover both structural and ligand field considerations [29] would suggest the $^6A_{1g}$ state to be more reasonable.

A $^6A_{1g}$ ground state gives rise to $^5E_{1g}$, $^5A_{1g}$ and $^5E_{2g}$ ion states. Evans et al. [49] assign the bands at 6.91, 10.10 and 10.51 eV respectively to these states. Further He(II) studies [48] confirm the high metal content of the orbitals giving rise to these bands. The ligand π-bands ionize around 8.72 eV; thus the spread of d-ionizations encompasses those of the top ligand levels. A worrying point though is the presence of a small shoulder at 11.10 eV assigned to C_5H_6 present as an impurity. A very weak band at 6.26 eV is believed to show the presence of a small amount of the $^2E_{2g}$ spin isomer [49].

The spectrum of $Mn(\eta\text{-}C_5H_4Me)_2$ shows a corresponding band of greater intensity [49]. The complex low energy region leads to the suggestion that this compound existed as a high spin $^6A_{1g}$, low spin $^2E_{2g}$ mixture in the gas phase; this hypothesis has been confirmed by structural studies [55]. The bands which correlate with those of unsubstituted manganocene are assigned in a similar manner. The extra bands corresponding to the low spin isomer are best considered with the spectrum of decamethylmanganocene.

---

[1] Recent He-II studies in the author's laboratory have confirmed this assignment

3. $^2E_{2g}$ Ground States

Polyalkyl substitution on manganocene appears to increase the ligand field splitting resulting in a low spin ground state for decamethylmanganocene. This molecule is isoelectronic with the ferricinium cation and shows the same $^2E_{2g}$ ground state $(e_{2g}^3 \, a_{1g}^2)$ rather than the $^2A_{1g}$ ground state $(e_{2g}^4 \, a_{1g}^1)$ found for most d^5 sandwich compounds of the early transition metals. The $^2E_{2g}$ molecular state gives rise to five ion states by one electron processes and these have all been observed and assigned by means of variations in intensity with photon energy. The ion state ordering is found to be $^3E_{2g} < {}^3A_{2g} < {}^1E_{1g} < {}^1A_{1g} < {}^1E_{2g}$ [48].

III. 16 Electron Compounds

Two different ground states are found for the 16 electron compounds. Bis-arene and mixed sandwich compounds show a diamagnetic $^1A_{1g}$ (e_2^4) ground state whereas the bis-cyclopentadienyl compounds are paramagnetic, the weight of the magnetic and P.E. evidence [56–58] suggesting a $^3E_{2g}(e_2^3 \, a_1^1)$ ground state. Some representative spectra are given in Fig. 4.

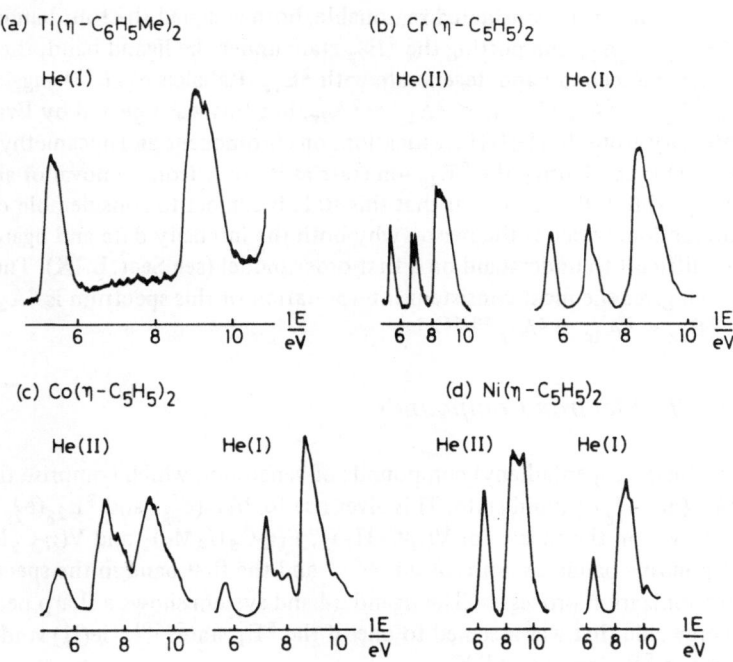

Fig. 4a–d. P.E. spectra of some 16, 19 and 20 electron compounds: **a** Ti(η-C$_6$H$_5$Me)$_2$ He(I); **b** Cr(η-C$_5$H$_5$)$_2$ He(I) and He(II); **c** Co(η-C$_5$H$_5$)$_2$ He(I) and He(II); **d** Ni(η-C$_5$H$_5$)$_2$ He(I) and He(II)

1. $^1A_{1g}$ Ground State

The e_{2g}^4 molecular configuration gives rise to just one ion state, $^2E_{2g}$, and just one d-band is observed in the P.E. spectrum of this class of compound [17,52,59]. The first I.E. for the compounds $Ti(\eta\text{-}C_6H_6)_2$ and $Ti(\eta\text{-}C_6H_5Me)_2$ is lower [59] than those of the mixed sandwich compounds $Ti(\eta\text{-}C_5H_5)(\eta\text{-}C_7H_7)$ [17] and $Zr(\eta\text{-}C_5H_5)(\eta\text{-}C_7H_7)$ [52], which is consistent with a higher ligand content of the e_2 orbital in the mixed sandwich compounds due to the more stable e_2 orbital on the isolated C_7H_7 ring.

2. $^3E_{2g}$ Ground State

Ionization of the $e_{2g}^3 a_{1g}^1$ configuration of chromocene gives rise to five ion states $^2E_{2g}(e_{2g}^3)$ and $^4A_{2g}$, $^2E_{1g}$, $^2A_{1g}$ and $^2A_{2g}(e_{2g}^2 a_{1g}^1)$. Four of these may be observed in the He(I) spectra [49,47,48] (Fig. 4). One band is clearly separated from the rest at low I.E. followed by a complex band in which at least three states are discernable. This spectrum posed considerable problems in assignment partly due to the unknown location of the fifth band and partly as no interpretation was consistent with both Cox's rules as to predicted band intensities and ligand field considerations. Cox et al. [57] originally assigned the first band to both $^2E_2$ and $^4A_2$ ionizations while higher bands were supposed to correspond to $^2E_1$, $^2A_1$ and $^2A_2$ ion states. However, this assignment was changed in a subsequent publication of Evans et al. [49]; two assignments were considered reasonable, both assigned the four bands to $^4A_{2g} < ^2E_{1g} < ^2A_{1g} < ^2A_{2g}$ one putting the $^2E_{2g}$ state under the ligand band, the other placing it under the second band degenerate with $^2E_{1g}$. Rabalais et al. [47] suggested the ordering $^2E_{2g} < ^4A_{2g} < ^2E_{1g} < ^2A_{1g} \sim ^2A_{2g}$, but this was rejected by Evans et al. [49] on intensity grounds. He(II) investigations on chromocene and decamethylchromocene [48] were able to identify the $^2E_{2g}$ ion state as it arises from removal of an electron from an a_1 orbital. It is also clear that this state is subject to considerable configuration interaction, which is the reason why both the intensity data and ligand field orderings are difficult to understand on a first order model (see Sect. B.IX). The assignment which gives the most consistent interpretation of this spectrum is $^4A_{2g} < ^2E_{1g} < ^2E_{2g} < ^2A_{1g} \sim ^2A_{2g}$ [48,56].

IV. 15 Electron Compounds

The bis-cyclopentadienyl compounds of vanadium, which comprise this class, have a $^4A_{1g}(e_{2g}^2 a_{1g}^1)$ ground state. This gives rise to $^3A_{1g}(e_{2g}^2)$ and $^3E_{2g}(e_{2g}^1 a_{1g}^1)$ ion states. However, in the spectra of $V(\eta\text{-}C_5H_5)_2$, $V(\eta\text{-}C_5H_4Me)_2$ and $V(\eta\text{-}C_5Me_5)_2$ the corresponding bands are unresolved [48,49], and the first band in the spectra is assigned to two ionization processes. The ligand π-band system shows a sharp peak on the low energy side that was assigned to one of the $^5E_1$ states [49]. He(II) studies suggest that it is the $^5E_{1u}$ ion band [48].

V. 19 Electron Compounds

The 19e compounds include cobaltocene and its derivatives which have a $^2E_{1g}(e_{2g}^4 a_{1g}^2 e_{1g}^1)$ ground state. The presence of so many d electrons, which give rise to seven ion states (see Table 3), and the increased nuclear charge to the right hand of the transition series leads to a complex P.E. spectrum with metal d and ligand π regions overlapping. Assignments here are less certain than for other sandwich compounds. The first band in all spectra, though, comes at a very low ionization energy, 4.71 eV for $Co(\eta$-$C_5Me_5)_2$, and may confidently be assigned to removal of the e_{1g} electron to give a $^1A_{1g}$ ion state [48,49]. This band shows an anomalously large intensity increase on changing from He(I) to He(II) radiation [48]. The second band is assigned to both $^3E_{1g}$ and $^3E_{2g}$ ion states. The third and fourth bands, clearly visible in the spectra of $Co(\eta$-$C_5H_5)_2$ and $Co(\eta$-$C_5H_4Me)_2$ [49], are presumably hidden under the ligand bands in the P.E. spectrum of $Co(\eta$-$C_5Me_5)_2$ [48]; a possible assignment is to the $^1E_{1g}$, $^1E_{2g}$ and $^3E_{1g}$ ion states. A small band at the high energy edge of the ligand band is then assigned to the remaining $^1E_{1g}$ ion state.

The iron compound $Fe(\eta$-$C_5H_5)(\eta$-$C_6Me_6)$ has a similar ground state to cobaltocene [60]. Its P.E. spectrum shows a low energy band with the same intensity characteristics as the first band of cobaltocene and its derivatives, and it is similarly assigned to a $^1A_{1g}$ state arising from removal of the antibonding e_1 electron. Three other d bands may be distinguished, but their intensity pattern differs from that found for cobaltocene [60]

VI. 20 Electron Compounds

This class comprises nickelocene and its substituted derivatives, which have $^3A_{2g}$ $(e_{2g}^4 a_{1g}^2 e_{1g}^2)$ ground states. Similar problems exist for the assignment of their P.E. spectra as were found for cobaltocene. The first band, which shows intensity characteristics similar to that of the cobaltocene e_{1g}^* band, is similarly assigned to removal of an e_{1g}^* electron to give a $^2E_{1g}$ ion state [47–49]. Such severe overlap exists between subsequent d bands and ligand bands that further assignment is difficult though He(II) studies do go some way to distinguishing the two types [48].

VII. Trends in Ionization Energies

Given such a large series of related compounds, we can distinguish regular variations in I.E. of corresponding bands. These will be interpreted with recourse to the assumption that deviations from Koopmans' theorem are constant for corresponding orbitals in a related series of compounds; they are therefore discussed in terms of the orbital energies from which the various types of electrons are ionized. We will discuss effects on the d ionizations and on the top ligand levels (e_{1u} or e_{1b} and e_{1g} or e_{1a}).

Table 4. Effect of methyl substitution on I.E. of sandwich compounds. Values give average shift (in eV) to lower I.E. relative to unsubstituted compound

Orbital	$M(C_5H_4Me)_2$	$M(C_6H_5Me)_2$	$M(C_6H_3Me_3)_2$	$M(C_5Me_5)_2$
a_{1g}	0.19	0.18	0.33	0.95
e_{2g}	0.18	0.28	0.54	0.88
e_{1g}^*	0.22			0.77
e_{1u}	0.26	0.36	0.77	1.25
e_{1g}	0.24		0.78	1.11

1. Variation with Ring Substituent

The effects of methyl substitution on the various I.E. bands are summarized in Table 4. Given the experimental error in measuring the I.E., which is ca. 0.05 eV for these sharp bands, the pattern in I.E. shifts is very constant. All shifts are to lower I.E. in comparison with unsubstituted derivatives, in line with the known inductive effect of the methyl group. The presence of two methyl groups in a metallocene has a fairly uniform effect on the d-ionizations, but decreases the ligand ionization energy slightly more. However, when two methyl groups are substituted in a bis-arene compound, the effect on the e_{2g} ionization is significantly greater than that on the a_{1g} ionization; this may reflect the higher ligand content of the e_{2g} orbital. Again the ligand bands show a slightly greater shift. The decamethylmetallocenes show a shift in their d and ligand π I.E. roughly five times that of the 1,1′-dimethylmetallocenes, whereas the bis-mesitylene compounds only show shifts roughly double those of the bis-toluene compounds.

2. Variation with Ring Size

The influence of ring size on metal ligand interactions have been commented on by various authors[17,28,52,61,62]. The predictions of simple qualitative M.O. arguments and semiempirical calculations are strikingly born out by the I.E. trends in sandwich compounds.

The principal bonding interactions in metal sandwich compounds are between the metal d_{xz} and d_{yz} orbitals and the e_1-π orbitals of the ring, and the metal d_{xy} and $d_{x^2-y^2}$ orbitals and the e_2-π ring orbitals. Moreover the e_1 bonding decreases and the e_2 bonding increases systematically with increasing size of the ligand rings. The $d_{z^2}(a_1)$ orbital is non-bonding.

Comparisons of I.E. are made most effectively between isoelectronic compounds, which necessitates a change in metal as well as in ring size along the series. Several plots of this type are shown in Fig. 5a, b. The a_1 I.E. increases rapidly across a series as the atomic number of the metal increases. As this orbital has largely metal character, this change can be correlated with a decrease of the metal d-orbital energy. The variation in e_2 I.E. is both less regular and less marked, which is understandable in terms of its composite nature. As far as the ring contribution to the e_2 orbital is concerned, the e_2 orbital should rise in energy across the series as the ring size de-

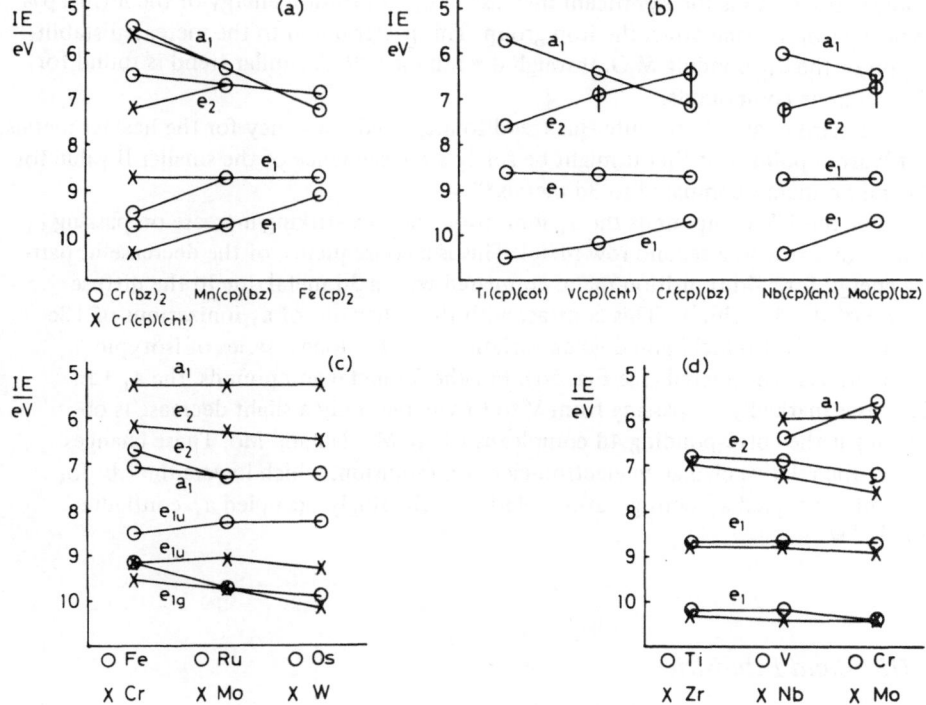

Fig. 5a–d. Variations in ionization energy in sandwich compounds: **a** 18 electron compounds, varying metal and ring; **b** 17 electron compounds, varying metal and ring (for the e_2 orbital, a weighted average of the $^3E_2$ and $^1E_2$ states is plotted; the bar denotes the exchange splitting; **c** 18 electron compounds, varying metal (bar indicates spin orbit splitting) ⊙ M(mecp)$_2$, X M(tol)$_2$; **d** 16, 17 and 18 electron isotypic compounds M(cp)(cht)

creases. Superimposed on this rise will be a decrease in orbital energy due to the metal contribution as the metal d-orbitals become more stable. The net effect is to produce an irregular variation. It is significant in this context that $Cr(\eta\text{-}C_5H_5)$-$(\eta\text{-}C_7H_7)$ has a higher e_2 I.E. than $Cr(\eta\text{-}C_6H_6)_2$ and $Nb(\eta\text{-}C_5H_5)(\eta\text{-}C_7H_7)$ has higher e_2 I.E.'s than $Nb(\eta\text{-}C_6H_6)_2$; the stabilizing effect of one C_7H_7 ring in this orbital outweighs that of two C_6H_6 rings. In the 17-e series the separation of $^3E_2$ and $^1E_2$ ion states decreases with ring size. This is also consistent with greater localization of the e_{2g} electrons on the metal for the smaller rings.

For the series $M(\eta\text{-}C_5H_5)(\eta\text{-}C_nH_n)$ the upper e_1 I.E. stays almost constant throughout whereas the lower e_1 I.E. decreases with decrease in ring size, following the increase in the ring orbital energy.

3. Variation with Metal

The trends in ionization energy caused by changing the metal within a particular subgroup are summarized graphically in Fig. 5c. For the 18e compounds the most

61

noteworthy trend is the significant increase in e_{1g} ionization energy of the $M(mecp)_2$ molecules on passing down the iron group. This is attributed to the increased stabilization of the e_{1g} bonding M.O. through d-π mixing [104]. A similar trend is found for the bis-arene compounds.

Orchard et al. [46] attribute this trend to increased covalency for the heavier metals, but Warren points out that it might be solely a consequence of the smaller B value for 4d and 5d metals compared to 3d metals [29].

For the 17e compounds the a_1 ionization shows a striking increase on passing from a first row to a second row metal. This is a consequence of the decrease in pairing energy for a 4d transition metal compared with a 3d metal due to the diffuse nature of the 4d orbitals. This contrast with the behaviour of a_1 ionizations of 18e compounds is also highlighted when variations in I.E. along a series of isotypic compounds is considered (see Fig. 5d). For the 3d metal compounds, the a_1 I.E. decreases markedly on passing from V to Cr whereas only a slight decrease is observed for the corresponding 4d complexes where M = Nb and Mo. These changes reflect the relative change in electron-electron repulsion, which lowers the I.E. for a doubly occupied a_1^2 configuration relative to the singly occupied a_1^1 configuration [17,52].

VIII. Band Intensities

Two types of assumption have been made in discussing band intensities in sandwich compounds. One is that M.O. cross sections may be obtained by summing the A.O. contribution to the M.O. weighted by the atomic cross-section (see Sect. A.IV). As atomic cross sections vary with photon energy, this model also provides a basis for discussion of the variation of relative band intensity on changing the ionizing radiation. The second is that for ionization from a particular subshell, relative band intensities are given by Cox's rules [27] (see Sect. A.V).

Overall the pattern of M.O. cross sections, as interpreted over a wide range of compounds by the former model, is consistent with the relative amounts of M(d) and C(2p) A.O. contribution to the M.O. as evidenced by ionization energy trends, exchange splittings and calculations. For example, the He(I) a_{1g}:e_{2g} band ratios for $Cr(\eta\text{-}C_6H_6)_2$ (1:4), $Mn(\eta\text{-}C_5H_5)(\eta\text{-}C_6H_6)$ (1:3) and $Fe(\eta\text{-}C_5H_5)_2$ (1:2:5) are consistent with the increasing metal content of the e_{2g} orbital across the series [28]. He(I): He(II) intensity changes have also been interpreted successfully along these lines [20,48]. There are, however, worrying exceptions where the model breaks down. For first row metallocenes ligand π-orbitals appear to have a larger cross section than the metal d-orbitals [46], whereas in $Os(\eta\text{-}C_5H_4Me)_2$ no substantial difference in cross section is discernable between the ligand $e_{1u}(\pi)$ and metal d orbitals; in consequence the a_{1g}:e_{2g} band ratio is almost exactly 1:2. Whereas the reduced intensity of the $e_{1g}(\pi)$ band relative to the non bonding $e_{1u}(\pi)$ band can be attributed to admixture of metal d A.O. in the case of iron, some additional effect must be operative in the case of osmium, where d and π orbitals have comparable cross section. Also, when considering intensity changes, it has been established empirically [48] that

$$\frac{k_{e_{1g}^{*}}^{II}}{k_{e_{1g}^{*}}^{I}} > \frac{k_{a_{1g}}^{II}}{k_{a_{1g}}^{I}} > \frac{k_{e_{2g}}^{II}}{k_{e_{2g}}^{I}} > \frac{k_{e_{1g}}^{II}}{k_{e_{1g}}^{I}} > \frac{k_{e_{1u}}^{II}}{k_{e_{1u}}^{I}} \ .$$

This order fits with the orbitals of higher d character having large $k_{\phi}^{II}/k_{\phi}^{I}$ ratios except in the case of the e_{1g}^{*} orbital, which must be presumed to have substantial C_{2p} character and therefore exhibits an anomalously high ratio.

Further studies of intensity changes over a wider range of photon energies are clearly desirable to establish the validity of using intensity variations to deduce the A.O. constituents of M.O. .

The sandwich compounds constitute the largest class of transition metal open shell compounds studied by UPS and thus provide a good test of the experimental validity of Cox's rules [27,57]. Obtaining intensity data is not always straightforward as bands due to different ion states frequently overlap; as the line shapes are unknown, they cannot be separated accurately. Within these limitations it has been found that the intensity rules give good predictions of relative intensities of ion states arising from the same configuration and consistent values for orbital cross sections when more than one configuration is involved. However, when the ion states produced are involved in significant configuration interaction (C.I.), as is found to be the case for chromocene and $^{2}E_{2}$ decamethylmanganocene, the intensity pattern does not concur with first order predictions and must be modified to take into account the proportion of the ion configuration, which may be reached by a one electron process, that contributes to the ion state [48].

IX. Ligand Field Treatments of Sandwich Compounds

The accessibility of several ion states on ionization of an open shell molecule enables the parameterization of the energies of these states with a ligand field model. The ligand field energy expressions and interaction matrices for sandwich compounds have been tabulated by Warren [29] and will not be reproduced here, though some first order energies are given in Table 3.

The orbital splitting pattern for sandwich compounds is defined by two parameters Δ_1 and Δ_2; B and C are Racah parameters and β the nephelauxetic ratio.

$$
\begin{array}{ll}
\underline{\qquad\qquad} & e_1, \pi \\
\quad\uparrow & \\
\Delta_1 & \\
\quad\downarrow & \\
\underline{\qquad\qquad} & a_1, \sigma \\
\quad\uparrow & \\
\Delta_2 & \\
\quad\downarrow & \\
\underline{\qquad\qquad} & e_2, \delta \ .
\end{array}
$$

A common assumption is that $C = 4B$. In crystal field theory, the d orbital energies are effectively core energies: the electron repulsions within the d-shell are not included. They thus differ significantly from S.C.F. orbital energies, which take into account all electron-electron repulsions.

Warren has analysed the P.E. spectra of $^2A_1$ d^5 and $^1A_1$ d^6 compounds. From the three d bands of the former ($^3E_2$, $^1E_2$ and $^1A_1$), values of B and Δ_2 are obtained and hence a value for β. Given a similar value of β, Δ_2 is then deduced for the isotypic d^6 compound. Values found are shown in Table 5. For Δ_2 one obtains the result Mcp$_2 \ll$ Mcpbz $<$ Mbz$_2 <$ Mcpch and for β values the contrary trend, Mcp$_2 \sim$ Mcpbz $>$ Mbz$_2 \gg$ Mcpch. These conclusions concur with those from the simple M.O. approach discussed above.

Table 5. Ligand field parameters deduced for sandwich ions [29]. Values are given in eV

Compound	B	β	Δ_2
$[V(mes)_2]^+$	0.047	0.64	1.78
$[Cr(mes)_2]^+$	0.052	(0.64)	1.91
$[Cr(cp)(bz)]^+$	0.059	0.73	1.66
$[Mn(cp)(bz)]^+$	0.064	(0.73)	1.64
$[V(cp)(cht)]^+$	0.032	0.43	1.75
$[Cr(cp)(cht)]^+$	0.035	(0.43)	2.29
$[Nb(cp)(cht)]^+$	0.024	0.45	2.19
$[Mo(cp)(cht)]^+$	0.026	(0.45)	2.19

The P.E. spectrum of decamethylmanganocene, which shows five ion states, provided a good test of the ligand field approach. It was found that a consistent fit could only be obtained when configuration interaction in the $^1A_1$ state was taken into account [48]. If C.I. was considered, agreement was excellent. Assumption of similar values for Δ_2, B and C for the other decamethylmetallocenes gave a good account of the other metallocene spectra [48], as shown in Table 6. The values fit well into the sequence deduced by Warren. However Warren's estimates neglect configuration interaction in the $^1A_1$ d^4 ion state; when taken into account, C.I. may modify his values slightly.

C. Lanthanide and Actinide Cyclooctatetraene and Cyclopentadienyl Compounds

I. Bis-cyclooctatetraene Actinides

The bis-cyclooctatetraene actinides, $M(\eta\text{-}C_8H_8)_2$, are structurally the closest analogues, for the 5f elements, of the d-block transition metal sandwich compounds. Their orbital structure (see Fig. 6) differs, however, in that the e_{2u} orbital is occupied by

Table 6. Energies of excited states of the decamethylmetallocene molecular ions calculated by ligand field theory with limited configuration interaction[a] [48]

$\{V(\eta\text{-}C_5Me_5)_2\}^+$		$^3A_{2g}$	$^3E_{2g}$				
Calculated		0	0.02				
Experimental		0	0				

$\{Cr(\eta\text{-}C_5Me_5)_2\}^+$	$^4A_{2g}$	$^2E_{1g}$	$^2E_{2g}$	$^2A_{1g}$	$^2A_{2g}$	$[^2E_{2g}]$	
Calculated	0	1.35	1.87	2.00	2.09	3.32	
Experimental	0	1.25	1.41	1.76	1.76		

$\{Mn(\eta\text{-}C_5Me_5)_2\}^+$	$^3E_{2g}$	$^3A_{2g}$	$^1E_{1g}$	$^1A_{1g}$	$^1E_{2g}$	$[^1A_{1g}]$	
Calculated	0	0.39	1.04	1.12	1.39	3.38	
Experimental	0	0.39	1.04	1.17	1.39		

$\{Fe(\eta\text{-}C_5Me_5)_2\}^+$	$^2E_{2g}$	$^2A_{1g}$					
Calculated	0	0.72					
Experimental	0	0.40					

$\{Co(\eta\text{-}C_5Me_5)_2\}^+$	$^1A_{1g}$	$^3E_{1g}$	$^3E_{2g}$	$^1E_{1g}$	$^1E_{2g}$	$^3E_{1g}$	$^1E_{1g}$
Calculated[b]	0	1.68	1.99	2.51	2.64	2.65	4.42
Experimental[c]	0	1.68	(2.3–4.3)				

$\{Ni(\eta\text{-}C_5Me_5)_2\}^+$	$^2E_{1g}$	$^4A_{2g}$	$^4E_{2g}$	$^2A_{2g}$	$[^2E_{2g}]$	$^2E_{2g}$	$[^2A_{2g}]$	$[^2E_{2g}]$
Calculated[b]	0	1.65	1.67	2.60	2.90	3.31	3.41	5.22
Experimental[c]	0	1.65		2.58	(1.2–3.2)			

[a] States in [ ] are those with minor components accessible by a one electron ionization of the ground state. Values calculated using B = 0.0925 C = 0.325 Δ_2 = 1.13

[b] Δ_1 = 2.67 (Co), 2.95 (Ni)

[c] Complex band includes ionization from e_{1u} and e_{1g} orbitals

four electrons. For the actinides the e_{2u} orbital can be a bonding level as the central atom possesses low lying 5f orbitals of e_{2u} symmetry that appear to have good overlap with the ligand $\pi\, e_{2u}$ combination [65]. Streitwieser proposes that this is a significant stabilizing force in these compounds [66].

He(I) and He(II) spectra of uranocene and thorocene are shown in Fig. 7, and the ionization energies are given in Table 7. An extra band at low I.E. in the uranocene spectrum is readily assigned to ionization of the f^2 configuration. This band undergoes a substantial increase in intensity on increasing the photon energy between He(I) and He(II) [24]. Similar increases are found for other 5f compounds [67,68]. Otherwise the spectra of the two molecules are very similar and have a common assignment. When only He(I) spectra were available [69,70], the first and second bands of the thorocene spectrum and the second and third of the uranocene spectrum were assigned to ionization from the e_{2g} and e_{2u} orbitals in accord with theoretical expectations at that time. Subsequent examination of intensity changes in the He(II) spectrum led to reversal of this assignment [24]. The lower I.E. band increases in intensity relative to the higher of these two bands on changing the ionization radiation from He(I) to He(II). This is assumed to be due to a 5f orbital contribution to the M.O. associated

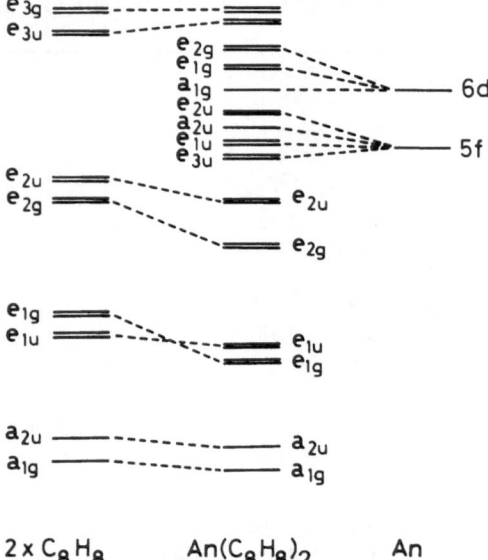

$$2 \times C_8H_8 \qquad An(C_8H_8)_2 \qquad An$$

Fig. 6. Interaction scheme for a bis-cyclooctatetraene actinide

with the lower ionization band, and so it is assigned to ionization from the e_{2u} orbital. This explanation of the intensity changes must be regarded as plausible rather than definitive in view of the anomalies found for the e_{1g}^{*} band of the metallocenes (see Sect. B.VIII). If it is assumed that differential relaxation for these orbitals is small, the implication is that the e_{2g} orbital lies lower than the e_{2u} orbital, which may in part be due to 6 d metal ligand bonding being greater than 5 f. Recent MS-Xα calculations [71] support this assignment and these conclusions.

(a) $Th(\eta - C_8H_8)_2$ (b) $U(\eta - C_8H_8)_2$

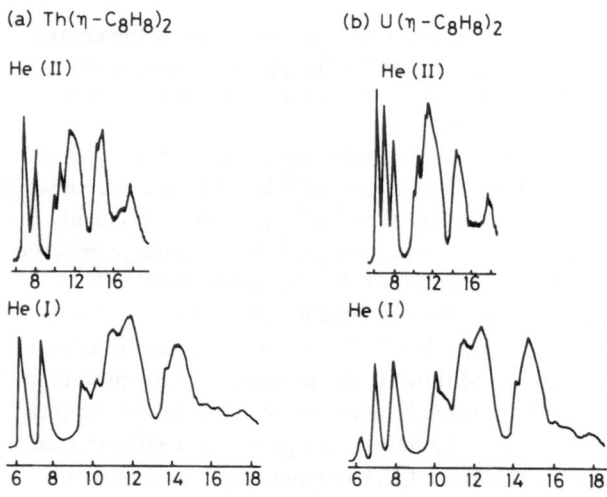

Fig. 7a, b. He(I) and He(II) spectra for thorocene and uranocene

Table 7. Ionization energies (eV) of some cyclopentadienyl and cyclooctatetraenyl lanthanide and actinide compounds

$M(\eta\text{-}C_8H_8)_2$

M	f^n	e_{2u}	e_{2g}	e_{1u}, e_{1g}	He(II)	References
Th		6.79	7.91	9.90, 10.14, 10.65	II	24, 69, 70
U	6.20	6.90	7.85	9.95, 10.28, 10.56	II	24, 69, 70

$M(\eta\text{-}C_5H_4Me)_3$

M		e_1 (π-mecp)				He(II)	References
Ce		7.37	7.88	8.51			72
Pr		7.28	7.86	8.41		II	68
Nd		7.29	7.86	8.42			72
Sm		7.16	7.85	8.31			72
Dy		7.06		8.39		II	68

$M(\eta\text{-}C_5H_5)_4$

M	f^n	e_1 (π-cp)				He(II)	References
Th		7.49	8.60	9.17		II	73
U	6.34	7.39	8.53	8.97		II	73

$U(\eta\text{-}C_5H_5)_3 \cdot THF$

f^n	e_1 (π-cp)			O lone pair		He(II)	References
6.42	7.58	8.18	8.73	10.03		II	73

The intensity ratios and the $e_{2u} - e_{2g}$ splitting, which is larger for thorium and uranium, both suggest that the ring e_{2u} $f_{\pm 2}$ mixing is greater for $U(\eta\text{-}C_8H_8)_2$ than for $Th(\eta\text{-}C_8H_8)_2$.

Subsequent bands between 9.8 and 11 eV are assigned to e_{1u} and e_{1g} ionizations, but their ordering is uncertain.

II. Cyclopentadienyl Compounds of f-Block Metals

The closest analogues of the d-block metallocenes, for the f-block elements that have been examined by P.E.S., are the tris- and tetra-cyclopentadienyl complexes, $M(\eta\text{-}C_5H_5)_3$ and $M(\eta\text{-}C_5H_5)_4$. For uranium the tris-cyclopentadienyl retains a donor ligand in the vapour phase and has been measured as $U(\eta\text{-}C_5H_5)_3 \cdot THF$.

1. $Ln(\eta\text{-}C_5H_4Me)_3$

He(I) spectra have been measured for $M(\eta\text{-}C_5H_4Me)_3$ where M = Ce, Pr, Nd, Sm and Dy [68,72], and He(II) where M = Pr and Dy [68]. The He(I) spectrum where M = Pr is presented in Fig. 8, and the ionization energies in Table 7. The spectra show a band system between 7 and 9 eV that is assigned to ionization of the upper e_1 π-levels of the

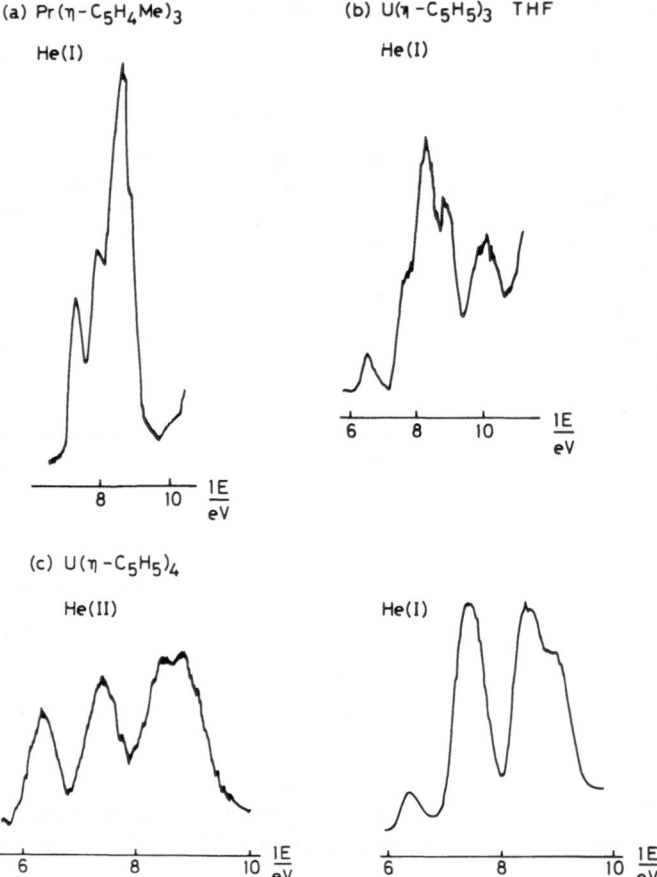

(a) Pr(η-C_5H_4Me)$_3$

He(I)

(b) U(η-C_5H_5)$_3$ THF

He(I)

(c) U(η-C_5H_5)$_4$

He(II)

He(I)

Fig. 8a–c. Photoelectron spectra for cyclopentadienyl compounds of lanthanides and actinides: a Pr(η-C_5H_4Me)$_3$, He(I); **b** U(η-C_5H_5)$_3$. THF, He(I); c U(η-C_5H_5)$_4$, He(I) and He(II)

cyclopentadienyl rings. If D_{3h} symmetry is assumed for these molecules as an approximation for the gas phase structure, the e_1 ring orbitals transform as a_2', a_2'', e' and e''. The a_2' orbital is expected to be highest on the basis of unfavourable ligand-ligand interactions. The first band is assigned to ionization from this orbital. Subsequent bands are not resolved, and further assignment is not possible. The first band in the dysprosium compound occurs 0.32 eV lower than the cerium compound: this is attributed to increasing ligand-ligand interaction as the metal ion becomes smaller. Comparison of the He(I) and He(II) spectra show that the relative intensities of the bands remain almost constant. No evidence was found for either f- or d-orbital covalency in these compounds in accord with conclusions from other physical measurements.

Furthermore, there was no evidence at all of a band anywhere in either the He(I) or He(II) spectra that could result from ionization of a $4f^n$ configuration. This was attributed to the intrinsically low ionization cross section of these orbitals at low photon energies.

2. $An(\eta\text{-}C_5H_5)_4$

In the crystalline state $U(\eta\text{-}C_5H_5)_4$ has S_4 symmetry [74]. If the local $M\text{-}C_5H_5$ symmetry is approximated as $C_{\infty v}$, the molecule may be treated as having T_d symmetry. The electronic structure is then given as

$$1\,a_1^2\ 1\,t_1^6\ 1\,e^4, 1\,t_2^6, 2\,t_1^6, f^n$$

where only the π-electrons of the cyclopentadienyl rings are considered and $n = 2$ for U and $n = 0$ for Th.

The spectrum of $U(\eta\text{-}C_5H_5)_4$ is shown in Fig. 8: the spectrum of $Th(\eta\text{-}C_5H_5)_4$ is very similar except that it lacks a band equivalent to the first band of the P.E. spectrum of the uranium derivative [73]. Ionization energy data for both compounds are given in Table 7. The first band in the spectrum of the uranium compound increases in intensity between He(I) and He(II) spectra and is assigned to ionization of f^2 electrons. Three bands are found for both compounds in the region ca. $7-9.5$ eV and are assigned to ionization of ring e_1 π orbitals. Consideration of intensities and intensity changes suggest an ordering $2\,t_1 < 1\,e < 1\,t_2$ for the ionization energies though this assignment cannot be regarded as definite. However, significant variations in the relative intensities of these orbitals suggest f-orbital covalency for these molecules.

3. $U(\eta\text{-}C_5H_5)_3 THF$

The spectrum of this compound is shown in Fig. 8, and its ionization energies are given in Table 7. Though the overall symmetry in the gas phase must be low, C_{3v} symmetry was assumed for analysis of the electronic structure of the $M(\eta\text{-}C_5H_5)_3$ unit. In this point group the Cp, e_1 $p\pi$ orbitals give rise to four symmetry adapted linear combinations transforming as $a_1, a_2, 2 \times e$. Ionizations from these orbitals are assigned to the three overlapping bands in the spectrum between 7.5 and 9 eV. A band at 10.3 eV is assigned to ionization of the O lone pair of THF, which is presumably forming the donor link to uranium as it occurs at a slightly higher I.E. than in free THF.

The low I.E. band at 6.43 eV is relatively more intense in the He(II) spectrum than in the He(I) spectrum and is assigned to ionizations from the f^3 configuration. It is a symmetrical band with no apparent splitting and a width at half-height of 0.5 eV.

III. f^n Ionizations

The final state structure expected on ionizing a 5 f sub-shell is influenced by two important factors. These are the nature of the appropriate atomic coupling scheme and the effects of the ligand field on the initial and final states. The first factor may be illustrated by consideration of the $^3H_4$ ground state for the U^{4+} ion. The energies of the final states $^2F_{7/2}$ and $^2F_{5/2}$ are independent of the choice of atomic coupling schemes as they are separated by $\frac{7}{2}\zeta_{5f}$, where ζ_{5f} is the spin-orbit coupling constant for the 5f sub-shell. On the other hand, the intensities of the final-state peaks are dependent on the mechanism of ground-state coupling.

Cox has treated the problem in the Russell-Saunders limit using fractional parentage methods[27] and predicts a value of 1.714:0.286 for the $^2F_{5/2}:^2F_{7/2}$ ratio. More recently the general intermediate coupling situation has been considered[75] and the value of 1.948:0.052 was predicted. In the j-j coupling limit only the $^2F_{5/2}$ state may be reached.

Ligand field perturbations on the ionic $5f^1$ states should be considered as well as the spin-orbit effects. Calculations on halides and borohydrides[67] show the ligand field effects to be weak, splittings of the J = 5/2 states being less than 0.2 eV.

In view of these general considerations, the observation of only one band for the ionization of the f^2 configuration in the uranium(IV) cyclopentadienyl derivatives is reasonable; the band is assigned to the $^2F_{5/2}$ ion state with any ligand field splitting unresolved.

In the case of a $5f^3$ configuration and $^4I_{9/2}$ ground term, the accessible final states are $^3H_4$, $^3H_5$ and $^3F_2$, and their relative final intensities in an intermediate coupling situation are predicted to be 2.137:0.187:0.612. Since the separation of the $^3H_4$ and $^3F_2$ state should be > 1 eV, at least two bands are expected in the P.E. spectrum of an f^3 ionization. Unfortunately the quality of the data on $U(\eta-C_5H_5)_3$-THF does not allow a definite conclusion as to the number of f-bands.

IV. Trends in Cyclopentadienyl e_1 Ionizations

Ranges for the vertical ionization energies of metal cyclopentadienyl compounds are summarized in Table 8. They show that the ionization energies of the ring e_1 orbitals decrease as the ionic character of the complexes increases. This trend can be readily interpreted in that the more ionic the character of the complex, the greater the negative charge on the cyclopentadienyl rings and the lower the ionization energies of the orbitals. Also if an e_1 orbital is involved in forming a covalent bond, the electrons occupying it are delocalized onto the metal and are consequently more difficult to ionize.

Table 8. Vertical Ionization Energies (eV) for ring e_1 orbitals in metal cyclopentadienyl compounds

Compound	I.E. range
$M(\eta-C_5H_5)_2$	8.4−10
$An(\eta-C_5H_5)_4$	7.4 − 9.2
$Ln(\eta-C_5H_4Me)_3$ [a]	7.0− 8.5

[a] A methyl substituent on a cyclopentadienyl ring consistently shifts e_1 ionizations by 0.2−0.3 eV

D. Bis-η-Cyclopentadienyl Metal Complexes

The $M(\eta\text{-}C_5H_5)_2$ moiety exists in such an extensive and varied number of compounds that its orbital structure has attracted much interest[76-81]. The stereochemistry of the $M(\eta\text{-}C_5H_5)_2$ fragment is quite constant, the two rings being non-parallel but η-bonded to the metal, which may be bound to one, two or three further groups. A useful fragment analysis for this class of molecule[77,78,81] is to start with the M.O. scheme for a metallocene unit with D_{5h} symmetry and to investigate the effects on the symmetry and energy of the orbitals of lowering the symmetry to C_{2v}. The resulting effect on the top three orbitals of ferrocene is shown in Fig. 9[81]. The three frontier orbitals of a $M(\eta\text{-}C_5H_5)_2$ unit are of $2 \times a_1 + b_1$ symmetry and they have a capacity to bind ligands in the xz plane (see Fig. 9)[2]. Ionization energy data is given in Table 9.

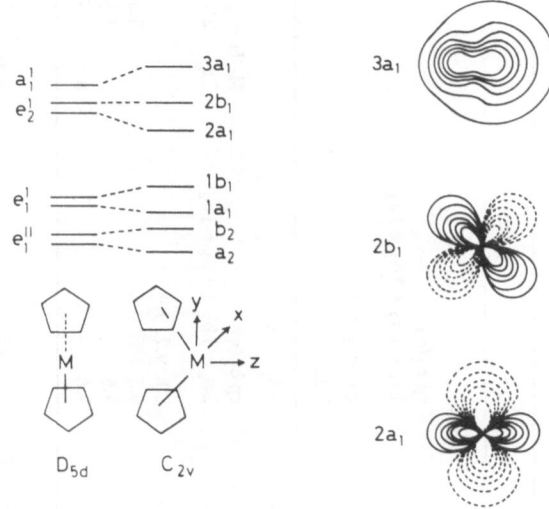

Fig. 9. Molecular orbitals for a bent, C_{2v}, $M(\eta\text{-}C_5H_5)_2$ fragment, adapted from [78] and [81]

I. $M(\eta\text{-}C_5H_5)_2(CO)_n$

The simplest P.E. spectra are those of the carbonyls, where the ligand ionization bands all occur above 14 eV. Figure 10 shows the low energy region for $Mo(\eta\text{-}C_5H_5)_2CO$ [78] and $Ti(\eta\text{-}C_5H_5)_2(CO)_2$ [82]. In the monocarbonyl the $Mo(\eta\text{-}C_5H_5)_2$ unit provides an empty acceptor orbital of $3\,a_1$ symmetry and a filled donor orbital of $2\,b_1$ symmetry to satisfy the bonding requirements of CO. Two low energy bands are observed, the first is assigned to the lone pair of electrons occupying the $3\,a_1$ orbital, a non-bonding orbital with high density along the x axis. This first band is relatively sharp, in line

---

2   A plethora of different coordinate systems have been used for these molecules, so care must be taken in comparing results of different papers

Table 9. Ionization energies of bis-cyclopentadienyl metal derivatives[a,b]

Bis-cyclopentadienyl metal carbonyls

	d-ionizations		cp $e_1(\pi)$	He-II	References
	$3a_1$	$2b_1$	$2a_1 + 1b_2 + 1b_1 + 1a_2$		
cp$_2$MoCO	5.9	6.8	8.8, 9.3, 9.6		78
cp$_2$Ti(CO)$_2$	6.35		9.15	II	82

Bis-cyclopentadienyl metal hydrides

	d-ionizations		cp $e_1(\pi)$	M–H σ			References
	$3a_1$	$2b_1$	$2a_1 + 1b_2 + 1b_1 + 1a_2$	$3a_1$	$2b_1$	$1a_1$	
cp$_2$ReH	6.4	7.0	8.8, 9.2, 9.9				78
cp$_2$MoH$_2$	6.4		9.5		8.9	c	78
cp$_2$WH$_2$	6.4		9.6		8.9	c	78
cp$_2$TaH$_3$			9.6	8.1	8.7	10.6	78

Biscyclopentadienyl metal halides[d]

	d-ionizations	cp $e_1(\pi)$	halogen p(π) (and M–X σ[e])	He-II	References
cp$_2$TiF$_2$ (A)		8.1, 8.7, 9.4	13.0	II	23
cp$_2$TiCl$_2$ (A)		8.5 (8.9), 9.1 (9.9)	10.2, 10.7, 11.1	II	23, 80, 85
Mecp$_2$TiCl$_2$		8.22, 8.65, 8.88, 9.46	10.03, 10.60, 11.06		80
cp$_2$ZrCl$_2$ (A)		8.6, 9.1, 9.8	10.5, 11.1, 11.3	II	23, 85
cp$_2$HfCl$_2$ (A)		8.5, 9.3, 10.0	10.6, 11.3, 11.6	II	23, 85
cp$_2$TiBr$_2$ (B)		9.6, 10.0, 10.5	8.8	II	23
cp$_2$ZrBr$_2$ (C)		9.9, 10.4, 10.8	8.9	II	23
cp$_2$TiI$_2$ (C)		9.6, 10.1	8.0, 8.3, 8.9, 9.3	II	23
cp$_2$ZrI$_2$ (C)		10.0	8.1, 8.4, 9.5	II	23
Mecp$_2$NbCl$_2$ (B)	6.4	9.8, 10.5, 10.8	8.6, 8.9	II	23
cp$_2$TaCl$_2$ (B)	6.4	10.3, 11.0	8.8, 9.2	II	23
cp$_2$TaBr$_2$ (C)	6.4	9.9, 10.5	8.8	II	23

Mecp$_2$MoCl$_2$ (C)	6.8	9.7, 10.3, 10.6	8.7, 8.9	II	23, 86
Mecp$_2$MoBr$_2$ (C)	6.9	9.6, 10.0	8.8	II	23, 86
Mecp$_2$MoI$_2$ (C)	6.8	9.5, 9.9, 10.4	7.6, 8.0, 8.1, 8.9	II	23, 86
cp$_2$VCl	6.8 (b$_1$), 7.42 (a$_1$)	9.47, 9.81	8.29, 10.35	II	83
cp$_2$VBr	6.8 (b$_1$), 7.43 (a$_1$)	9.37, 10.07	8.14, 8.90	II	83
cp$_2$VI^f	6.71 (b$_1$), 7.33 (a$_1$)	9.02, 9.34	7.69, 8.30	II	83

[a] Where the cp$_2$M unit is bound to an alkyl, olefin or allyl group, the I.E. data is given in the Tables associated with that ligand

[b] Symmetry labelling and orbital type is explained in figure and in the text

[c] 1 a$_1$ ionization assumed to overlap with the cp e$_1$(π) band

[d] Considerable cp e$_1$(π), halogen p(π) mixing is found for these compounds. Bands are classified according to the predominant A.O. character and the compound classification A, B or C is given, see text and reference[23]

[e] The M−X(σ) ionizations are difficult to distinguish and are assumed to lie with the X pπ bands

[f] Spin-orbit splitting shown on I(pπ) band

73

J.C. Green

(a) Mo(η-C₅H₅)₂CO

He(I)

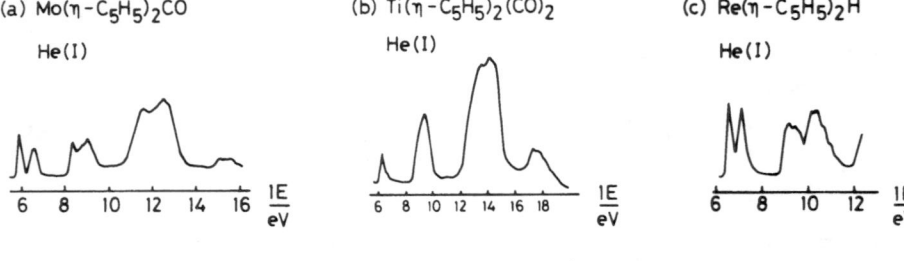

(b) Ti(η-C₅H₅)₂(CO)₂

He(I)

(c) Re(η-C₅H₅)₂H

He(I)

(d) W(η-C₅H₅)₂H₂

He(I)

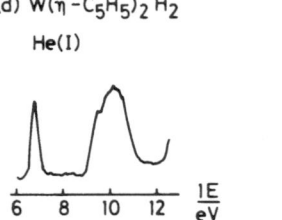

(e) Ta(η-C₅H₅)₂H₃

He(I)

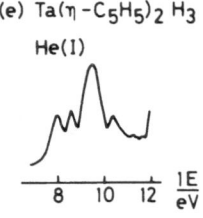

Fig. 10a–e. He(I) photoelectron spectra of: **a** Mo(η-C₅H₅)₂CO; **b** Ti(η-C₅H₅)₂(CO)₂; **c** Re(η-C₅H₅)₂H; **d** W(η-C₅H₅)₂H₂; **e** Ta(η-C₅H₅)₂H₃

with ionization from a non bonding orbital. The second band is broader and is assigned to the $2 b_1$ ionization, which is expected to exhibit bonding character due to back-donation to the CO $2 \pi^*$ orbital. Subsequent complex bands are assigned to ionization of orbitals of mainly cyclopentadienyl $e_1(\pi)$ character. On bending the $M(\eta-C_5H_5)_2$ fragment, the e_{1u} and e_{1g} orbitals adopt $a_1 + b_1 + a_2 + b_2$ symmetry.

The dicarbonyl $Ti(\eta-C_5H_5)_2(CO)_2$ shows a similar P.E. spectrum [82] but in this case there is only one d-band, which is associated with ionization from an a_1 orbital involved in back donation to the two CO groups.

II. $M(\eta-C_5H_5)_2H_n$

The isoelectronic series of hydrides $Re(\eta-C_5H_5)_2H$, $M(\eta-C_5H_5)_2H_2$, where M = Mo and W, and $Ta(\eta-C_5H_5)_2H_3$ clearly show the presence of two, one and no lone pairs respectively in their photoelectron spectra [78]. This is a consequence of the hydrogen ligands providing symmetry-adapted combinations of H(1 s) orbitals, $a_1(H_1)$, $a_1 + b_1(H_2)$ and $2 a_1 + b_1(H_3)$, which match the frontier orbitals of the $M(\eta-C_5H_5)_2$ unit. A M.O. correlation diagram is shown in Fig. 11, and the spectra are illustrated in Fig. 10. The M–H σ ionizations are more difficult to identify, but in the case of $Ta(\eta-C_5H_5)_2H_3$, where the C_5H_5 $e_1(\pi)$ band appears very narrow, three other bonding bands are observed that may be assigned to the three M–H σ bonding orbitals.

74

III. $M(\eta\text{-}C_5H_5)_2X_n$

The bis-cyclopentadienyl metal dihalides have more complex spectra because the halogen p ionizations occur in the same region as the $C_5H_5\,e_1(\pi)$ bands. The pπ orbitals of the X_2 unit have identical transformation properties to the $C_5H_5\,e_1(\pi)$ orbitals, so mixing of the two sets is allowed. Relative intensity studies of He(I) and He(II) spectra have been of great value in elucidating the relative ordering of ionizations and the degree of mixing as Cl, Br and I np ionizations drop in intensity relative to C 2p ionization on increasing the photon energy. The compounds, which are known for a variety of halogens and group IVa, Va and VIa metals, may be roughly divided into three classes: class A where the $C_5H_5\,e_1(\pi)$ ionization occurs at lower energy than the halogen ionizations, class B where little relative intensity change is observed and the M.O. may be assumed to be of mixed character, and class C where the halogen pπ ionization bands lie below the $C_5H_5\,e_1(\pi)$ bands. Representative spectra are shown in Fig. 12.

Two factors are found to affect the class of a compound. As the electronegativity of the halogen decreases, a transition takes place from class A, through class B to class C. A similar transformation occurs on moving from left to right across the transition series. For example, for $M(\eta\text{-}C_5H_5)_2Cl_2$, the zirconium compound is class A, the niobium class B and the molybdenum class C. In these variations the key factor appears to be the electropositive nature of the metal. This is discussed further in Sect. D. IV, where analogous lanthanide and actinide complexes are considered.

P.E. spectra have been obtained for $V(\eta\text{-}C_5H_5)_2X$ where $X = Cl$, Br, I [83]. The paramagnetic nature of these d^2 compounds is reflected in the resolution of two d-bands arising from ionization of the singly occupied b_1 and a_1 orbitals. In all three cases the halogen pπ band occurs at lower energy than the cp $e_1(\pi)$ band. The spectrum of $V(\eta\text{-}C_5H_5)_2Cl_2$ published by Petersen et al.[80] shows two low energy

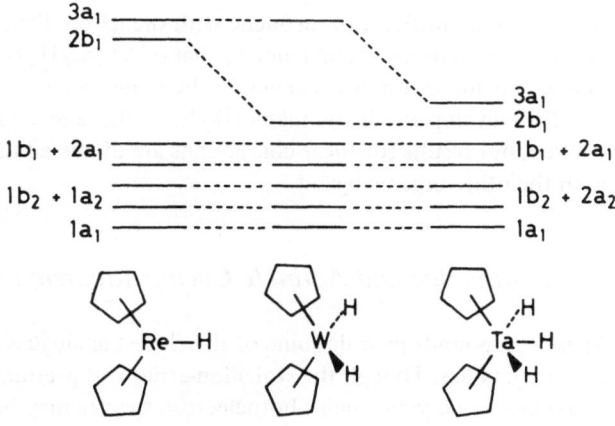

Fig. 11. Correlation diagram for $Re(\eta\text{-}C_5H_5)_2H$, $W(\eta\text{-}C_5H_5)_2H_2$ and $Ta(\eta\text{-}C_5H_5)_2H_3$

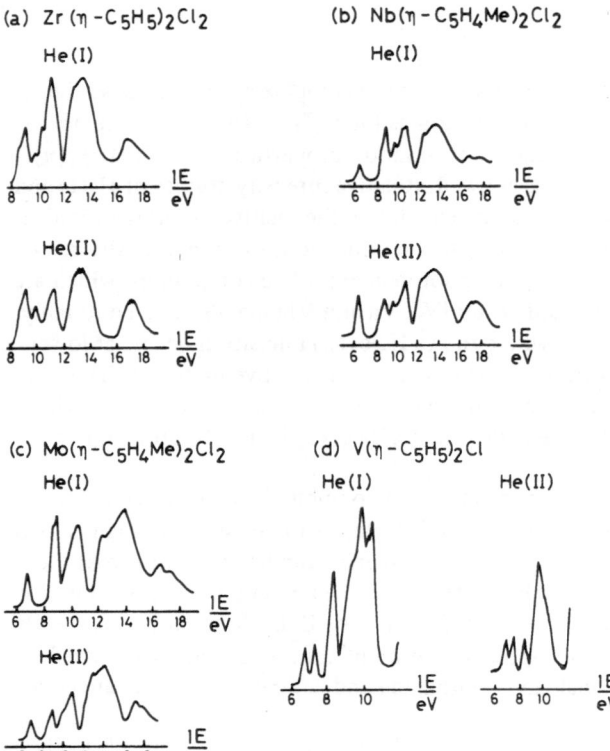

Fig. 12a–d. Photoelectron spectra of some bis-η-cyclopentadienyl metal halides: a $Zr(\eta\text{-}C_5H_5)_2Cl_2$; b $Nb(\eta\text{-}C_5H_4Me)_2Cl_2$; c $Mo(\eta\text{-}C_5H_4Me)_2Cl_2$; d $V(\eta\text{-}C_5H_5)_2Cl$

bands that are difficult to reconcile with the d^1 configuration. However, the spectrum shows a strong resemblance to that of $V(\eta\text{-}C_5H_5)_2Cl$, and so it is likely that *in situ* decomposition is occurring to the monohalide.

Bis-η-cyclopentadienyl metal alkyls, olefins and allyl complexes have also been studied, but results for these compounds are discussed under the sections associated with the other organic ligand.

IV. Lanthanide and Actinide Cyclopentadienyl Halides

These compounds provide some of the closest analogues to the transition metal organometallics. Though the stoichiometries and presumed symmetries of the gas phase molecules vary, their photoelectron spectra may be analysed by a common localized bond model. The range of compounds studied and their ionization energies are given in Table 10, and some representative spectra in Fig. 13.

Table 10. Ionization energies (eV) of lanthanide and actinide cyclopentadienyl metal halides and borohydrides

	f^n	cp e_1 (π)	halogen p	References
Y(η-C$_5$H$_5$)$_2$Cl		8.62	11.31	68
Gd(η-C$_5$H$_5$)$_2$Cl		8.59	11.58	68
Th(η-C$_5$H$_5$)$_3$Cl[a]		8.02, 8.61, 9.31	10.74	68, 88
Th(η-C$_5$H$_4$Me)$_3$Cl		7.75, 8.25, 8.95	10.45 (10.65)	68, 88
U(η-C$_5$H$_5$)$_3$Cl[a]	7.01	7.99, 8.56, 9.20	10.48	68, 88
U(η-C$_5$H$_4$Me)$_3$Cl[a]	6.91	7.92, 8.32, 8.95	10.41	68, 88
U(η-C$_5$H$_4$Me)$_3$Br	6.95	8.20, 8.50, 9.20	9.95	88
U(η-C$_5$H$_4$Me)$_3$BH$_4$	6.35, 6.75	8.10, 8.45, 8.85	10.30[b]	88

[a] Values taken from reference
[b] Band assigned to B-H ionization

A direct comparison may be made between the spectrum of the compounds M(η-C$_5$H$_5$)$_2$Cl where M = Y, Gd [87] and that where M = V discussed above, though it must be remembered that the yttrium and gadolinium compounds may be dimeric in the gas phase. The first notable difference is that for the gadolinium compound no f^n ionization process is observed, whereas the d ionizations are clearly evident for the vanadium compound. Secondly, in the "lanthanide" compounds the first complex band is due to ionization of the cp e_1 (π) orbitals and the halogen ionization occurs at much higher energy ($\sim$ 11.5 eV): this contrasts strongly with the vanadium compound where the halogen ionizations occur at 8.3 eV, below the cp e_1 (π) band. Thus the ionic lanthanide compounds belong to class A whereas the vanadium compound belongs to class C.

The triscyclopentadienylactinide halides also seem to be members of class A, in all cases the cp e_1 (π) ionizations occurring well below the halogen bands. In the case of the uranium compounds, however, ionization of the f-electrons is clearly visible as the first band: the f-band undergoes a large increase in relative intensity in the He(II) spectra [68,88].

The compound U(η-C$_5$H$_5$)$_3$BH$_4$ is unusual in that it appears to show two low energy bands attributable to f-ionizations, though no He(II) studies are reported on this compound [88].

Suggestions may be made for more detailed assignment of the cp bands to various symmetry adapted M.O., but they are largely based on energy expectations for the M.O. concerned and therefore must be regarded as tentative.

The bis-pentamethylcyclopendienyl actinidedihalides provide analogues of the bis-cyclopentadienyl transition metal dihalides. They resemble the compounds of zirconium and hafnium more than those of molybdenum in that they are class A [84] and indeed have higher halogen ionization energies, which again correlates with their greater ionicity. It is interesting that the chemistry of these compounds resembles that of the zirconium and hafnium analogues.

Though the trend to lower I.E. of the cyclopentadienyl e_1 ionizations with electropositive nature of the metal, which was discussed in Sect. C. IV, is readily understandable in terms of the negative charge carried by the organic group, the

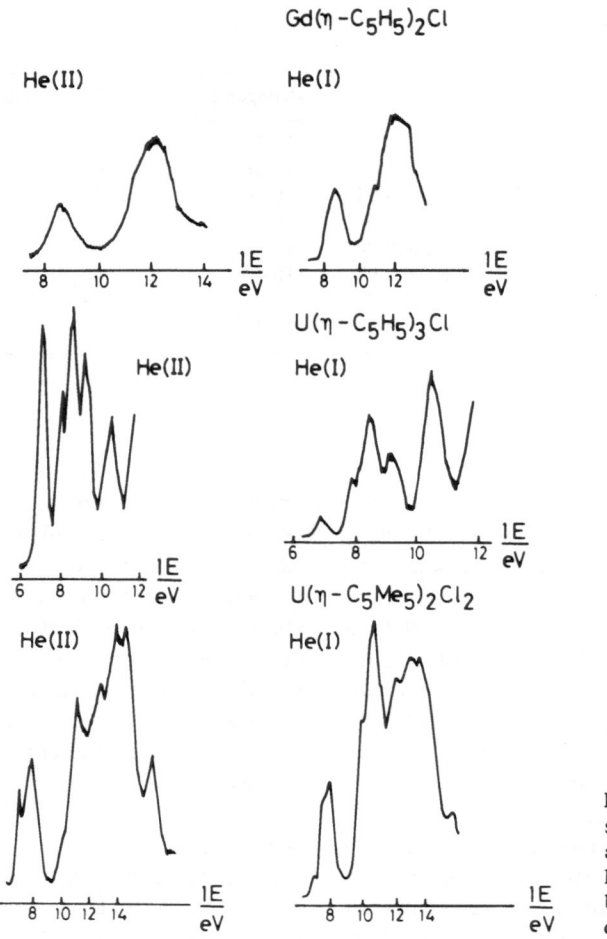

Fig. 13a–c. Photoelectron spectra of lanthanide and actinide cyclopentadienyl halides: **a** $Gd(\eta\text{-}C_5H_5)_2Cl$; **b** $U(\eta\text{-}C_5H_5)_3Cl$; **c** $U(\eta\text{-}C_5Me_5)_2Cl_2$

reverse trend of halogen ionization to higher I.E. observed for these molecules is much more puzzling and deserves further study. It may well be that the Madelung term is dominant in this case.

E. Metal Alkyls

The more convenient description of the bonding in this class of molecules is in terms of a framework of σ-bonds binding neighbouring atoms, *i.e.* a localized bond description. This has been recognized in deducing molecular orbital schemes necessary for assigning photoelectron spectra in so far as localized bond orbitals are used as basis sets. As in many cases localized bond orbitals appear to have characteristic ionization energies; this approach is of considerable assistance in spectral assignment[89].

I. Tetrahedral Metal Alkyls

Photoelectron studies have been made on the compounds $M(CH_2 CMe_3)_4$ and $M'(CH_2 SiMe_3)_4$ where $M = M' = $ Ti, Zr, Hf [91] and Cr [90]. Comparison was also carried out with main-group analogues where M = Ge and Sn [91] and M' = Ge [91], Sn [90,91] and Pb [90]. Typical spectra are shown in Fig. 14, and ionization energies given in Table 11.

In all spectra the higher energy region (above 10 eV for $M(CH_2 CMe_3)_4$ and above 9.5 eV for $M(CH_2 SiMe_3)_4$) closely resembles that of the parent hydrocarbon, neopentane or tetramethylsilane and is readily assigned by analogy[89] to bond ionization ($\sigma(C{-}H)$, $\sigma(C{-}C)$ and $\sigma(Si{-}C)$). In the d^0 compounds there is one more band at low energy that is assigned to $\sigma(M{-}C)$ ionization. If the local symmetry around the metal atom is taken as T_d the $\sigma(M{-}C)$ orbitals transform as $a_1 + t_2$, the assumed energy ordering being $t_2(\sigma(M{-}C)) > a_1(\sigma(M{-}C))$. The low energy band is assigned to the $t_2(\sigma(M{-}C))$ ionization while the $a_1(\sigma(M{-}C))$ ionization is assumed hidden under the ligand ionizations. In the chromium compound there are two bands in the low energy region; the second one is similarly assigned to the $t_2(\sigma(M{-}C))$ ionization, while the first band is due to ionization of the d-electrons, which occupy an orbital of e symmetry.

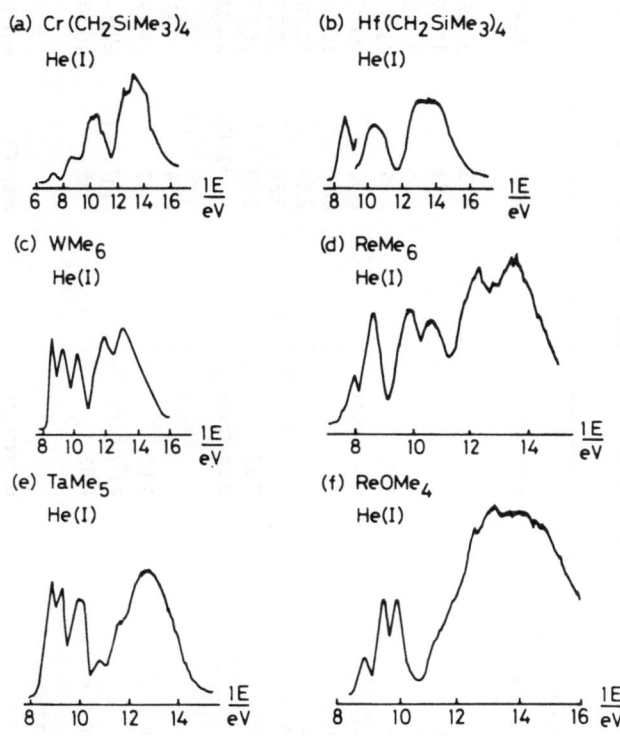

Fig. 14a–f. He(I) photoelectron spectra of transition metal alkyls and oxoalkyls:
a $Cr(CH_2SiMe_3)_4$; b $Hf(CH_2SiMe_3)_4$; c WMe_6; d $ReMe_6$; e $TaMe_5$; f $ReOMe_4$

Table 11. Ionization energies (eV) of metal-alkyl compounds

	d-electrons	σ(M–C)	(C–C)	(C–H)	He-II	References
Binary metal alkyls						
M(CH₂CMe₃)₄		σ(M–C)	(C–C)	(C–H)		
M =						
Ti		8.33	11.35	12.59		91
Zr		8.33	11.28	12.50		91
Hf		8.51	11.40	12.54		91
Cr	7.25	8.37	11.0	12.2		90
M(CH₂SiMe₃)₄	d-electrons	σ(M–C)	(Si–C)	(C–H)		
Ti		8.58	10.46	13.35		91
Zr		8.64	10.28	13.27		91
Hf		8.58	10.27	13.36		91
Cr	7.25	8.69	10.4	13.6		90
MMe₆	d-electrons	t_{1u}(M–C)	a_{1g}(M–C)	e_g(M–C)		
M =						
W	7.89	8.59	9.33	10.17	II	92, 93
Re		8.47	9.77	10.48		93
MMe₅	a_2''(M–C)	a_1'(M–C)	e'(M–C)	a_1'(M–C)		
Ta	8.83	9.25	10.10	11.14, 13.5, 13.9 II		92, 93
Metal oxo alkyls	d-electrons	e(M–C)	b_1(M–C)	(Si–C)		
ReOMe₄	8.86	9.5	9.9			93
ReO(CH₂SiMe₃)₄	8.00	8.46	8.93	10.2		93
Metal carbonyl alkyls						
M(CO)₅Me	e(M)	b_2(M)	a_1(M–C)	e(CH₃)		
Mn	8.65	9.12	9.49	12.6		95, 96
Re	8.71, 8.93	9.51	9.51	12.8		97, 98
Mn(CO)₅CF₃	9.17	9.51	10.53			95, 96
Metal carbonyl acyls						
M(CO)₅COCF₃	e(M)	b_2(M)	a_1(M–C)			
Mn		9.0ᵃ				96
Re	9.40, 9.69	9.97	8.80			98

Metal cyclopentadienyl carbonyl alkyls

	d orbitals	M–C	$C_5H_5e_1$	
Mo(η-C_5H_5) (CO)$_3$Me	7.78	9.07	9.7, 10.0	106
W(η-C_5H_5) (CO)$_3$Me	7.6, 7.77	9.26	9.92, 10.2	106
Fe(η-C_5H_5) (CO)$_2$Me	7.78, 8.53	9.15	9.90	106–108
Ru(η-C_5H_5) (CO)$_2$Me	8.13, 8.29, 8.96	9.48	9.98, 10.51	106
Fe(η-C_5H_5) (CO)$_2$CH$_2$CN	8.29, 8.90, 9.48	11.14	10.25	106

Titanium (IV) methyl compounds

	σ(M–C)	substituent orbitals	
TiMeCl$_3$	10.8	11.7, 12.7, 13.1,	109
TiMe(O-i-Pr)$_3$	9.8, 10.4^b	9.4, 9.8, 10.4	109
TiMe(η-C_5H_5) (OEt)$_2$	10.2	8.7, 10.2	109
TiMe(NEt$_2$)$_3$	10.1	7.6, 10.1	109

Platinum methyl tertiary phosphine complexes

	d-orbitals	σ(Pt–Me)	σ(Pt–P)^c	substituent orbitals^c	
trans-PtMeI(PMe$_3$)$_2$	7.33, 7.80, 8.23, 8.45	9.23		9.64, 11.64	105
trans-PtMeCl(PMe$_3$)$_2$	7.76, 8.20, 8.48, 8.73	9.29		10.07, 11.85	105
cis-PtMe$_2$(PMe$_3$)$_2$	7.68, 8.08, 9.36	9.13	9.99, 11.50		105
trans-PtMeI(PMe$_2$Ph)$_2$^{d}	7.12, 7.51, 8.22			9.28	105
trans-PtMeBr(PMe$_2$Ph)$_2$^{d}	7.43, 7.81, 8.08			9.18	105
trans-PtMeCl(PMe$_2$Ph)$_2$^{d}	7.54, 8.21			9.21	105
cis-PtMe$_2$(PMe$_2$Ph)$_2$^{d}	7.43, 7.99, 8.24	9.13		9.28	105

Bis-η-cyclopentadienyl metal alkyls

	d-orbitals 3a$_1$	σ(M–C) 2b$_1$ 1a$_1$	π-C_5H_5	
Mo(η-C_5H_5)$_2$Me$_2$	6.1	8.3 9.6	8.9	78
W(η-C_5H_5)$_2$Me$_2$	6.0	8.3 9.6	8.8, 9.0	78

a   Broad band
b   Considerable mixing between σ(M–C) and O orbitals
c   Considerable overlap of bands occurred: assignments are tentative
d   Bands obscured by phenyl bands

Interestingly, in the case of the compounds $M(CH_2 CMe_3)_4$, a high energy shoulder is observed on the first band in the case of M = Hf and a broadening in the case of M = Zr. Jahn-Teller splitting and spin orbit effects are rightly rejected as a cause of this asymmetry, and the authors attribute it to a D_{2d} distortion of the molecular ground state[91], citing support from vibrational studies. Another possible cause is ionization of the $a_1(\sigma(M-C))$ orbital.

The vertical $\sigma(M-C)$ I.E. of the trimethylsilylmethyl compounds are consistently slightly higher than for the neopentyl compounds, in accordance with the higher electron releasing properties of the latter ligand. For the group IVA compounds the first I.E. is insensitive to the nature of the central metal, whereas for the Group IVB compounds the expected decrease in first I.E. due to progressive decrease of electronegativity with atomic number, is observed.

One of the reasons cited for studying these compounds was to understand more about their high thermal stability. Both sets of authors conclude that the stability trends are not due to ground state electronic effects.

II. Metal Methyls

Spectra have been obtained for WMe_6 [92,93], $ReMe_6$ [93] and $TaMe_5$ [92,93]. An early report on WMe_6 [94] was incorrect [92]. Spectra are shown in Fig. 14, and ionization energies are given in Table 11.

For WMe_6 three bands are observed in the low I.E. region; they can be assigned in O_h symmetry to t_{1u}, e_g and a_{1g} $\sigma(M-C)$ ionizations, but the band intensities in no way correspond with orbital degeneracies. $ReMe_6$ shows an additional low energy band due to the d-electron. He(I)/He(II) intensity ratios and ionization energy trends lead to the I.E. assignment $3 t_{1u} < 2 a_{1g} < 2 e_g$.

$TaMe_5$ is assumed to have D_{3h} symmetry giving $2 a_1' + a_2'' + e'$ symmetry for the M-C bonding orbitals. A corresponding four bands are observed in the low energy region; they are tentatively assigned the ionization energy ordering $a_2'' < a_1' < e' < a_1'$.

These assignments are consistent with a view of the metal-σ-carbon bonding in transition metal alkyls in which the major metal component comes from the d-orbitals. There is also an appreciable binding contribution from W or Re 6 s orbitals, but very little from the 6 p orbitals.

III. Metal Oxoalkyls

Two rhenium oxoalkyls, $ReOR_4$ where R = Me and $CH_2 SiMe_3$, have been studied [93]. They are assumed to have a square pyramidal structure. The first band in both compounds is readily assigned to ionization of the d electron, which is assumed to occupy a b_2 orbital. The I.E. is about 1 eV higher for $ReOMe_4$ than $ReMe_6$. Only two bands are observed in the region where $\sigma(M-C)$ ionizations might be expected, though three bands are predicted. These are assigned to the e and b_1 ionizations, the a_1 ionization is assumed to lie under the ligand bands. The spectrum of $ReO(CH_2 SiMe_3)_4$ shows a broad band at 10.2 eV, which is assigned to Si-C bonding orbitals.

IV. Metal Carbonyl Alkyls and Acyls

The compounds studied in this class constitute derivatives of the $M(CO)_5$ group where M = Mn and Re. With respect to the alkyls and acyls four papers report experimental results [95-98], but there have been many attempts at spectral assignment [99-102]. Indeed perusal of these papers and others concerned with the assignment of the closely related halide and hydride derivatives provides a cautionary tale in the assignment of photoelectron spectra. In this review we restrict ourselves to discussing the evidence for what is now generally accepted as the correct assignments of the P.E. spectra of these compounds.

Spectra of $Mn(CO)_5Me$, $Re(CO)_5Me$ and $Mn(CO)_5CF_3$ are shown in Fig. 15, and ionization energies of these compounds together with those of $Mn(CO)_5 COCF_3$ and $Re(CO)_5COCF_3$ in Table 11.

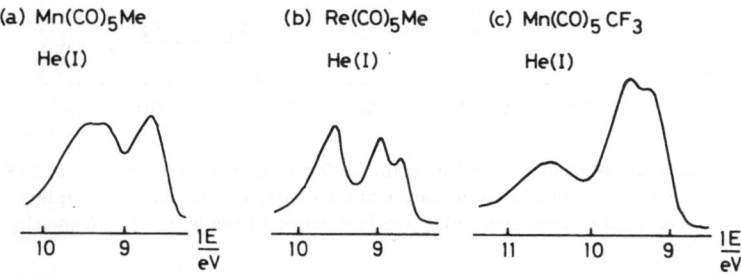

Fig. 15a–c. He(I) photoelectron spectra of metal pentacarbonyl alkyls: a $Mn(CO)_5CH_3$; b $Re(CO)_5CH_3$; c $Mn(CO)_5CF_3$

If the $M(CO)_5$ group is treated as an octahedral fragment (C_{4v} symmetry), it is readily shown that the t_{2g} orbitals now transform as $e(d_{xz}, d_{yz})$ and $b_2(d_{xy})$ and that the fragment provides an orbital of a_1 symmetry which can bind a further ligand [103]. The M.O. sequence $b_2 < e$ is also anticipated on the grounds that the d_{xz} and d_{yz} orbitals now interact less with the CO π^* orbitals than does the d_{xy} orbital; however, this ordering could be changed by interaction with a ligand. In the case of the rhenium compounds the ^2E ion state of $Re(CO)_5X$ will be split by spin-orbit coupling. This leads to three possible energy level schemes [98] (see Fig. 16).

The P.E. spectrum of $Re(CO)_5H$ shows four bands in the low energy region. The relative spacings of the first three bands are only consistent with scheme 3, and they are assigned an ionization energy ordering $e''(e) < e'(e) < e''(b_2)$; the fourth band is assigned to the a_1(Re-H) ionization. In the case of $Mn(CO)_5H$, no spin-orbit splitting of the e band is observed but otherwise the assignment is similar.

The $M(CO)_5Me$ compounds are assigned partly by analogy. In the rhenium compound only three bands are clearly defined, but the highest band has a hint of a high energy shoulder; for this reason, and on intensity grounds, it is assumed to

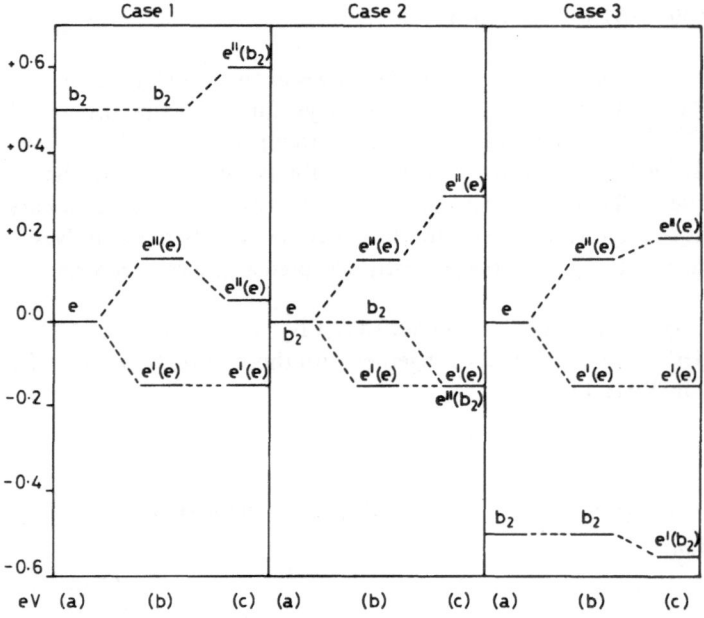

Fig. 16 a–c. Molecular orbital diagram for three possible cases of spin-orbit (S.O.) coupling in ML(CO)$_3$ systems. (a) represents e and b$_2$ energies with no S.O. coupling, (b) shows splitting of e level by S.O. coupling, (c) shows interaction of the b$_2$ level with the e''(e) level

result from two ionization processes. Again the splitting pattern is only consistent with case 3, and the ionization energy assignment e''(e) < e'(e) < e''(b$_2$) ~ e'(a$_1$) follows. The manganese compound shows three bands, the second and third being very close, and, by analogy with the rhenium case, the assignment of ionization energy bands e < b$_2$ < a$_1$ is given.

In both cases the a$_1$ ionization lies about 1 eV lower in the methyl compounds than in the analogous hydrides; also the e$_2$ − b$_2$ separation is about 0.2 eV larger in the methyl than the hydride. This latter fact is taken as evidence for a hyperconjugate effect in the methylated compounds, the e orbitals on the methyl group interacting with and destabilizing the e orbitals of the metal.

For Mn(CO)$_5$CF$_3$ a similar ionization energy ordering is assumed, but now the a$_1$ band (10.53 eV) is well separated from the b$_2$ band (9.51 eV), owing to the greater electronegativity of the CF$_3$ group.

The spectrum of Re(CO)$_5$COCF$_3$ shows four bands in the low energy region. In this case the first band is assigned to the a$_1$ ionization and subsequent ones to e''(e) < e'(e) < e''(b$_2$), as they closely resemble the splitting pattern of Re(CO)$_5$H. That the σ-bond has a lower ionization energy than the metal d-orbital is attributed to the a$_1$ orbital possessing C–O antibonding character[98]. In the case of the manganese analogue, the spectrum is too diffuse to assign.

Related silyl and germyl compounds have also been studied[104].

V. Other Metal Alkyls

Various other compounds containing metal alkyl bonds have been studied by He(I) P.E.S., and in most cases reasonable assignments of the M–C ionization have proved possible. In the case of some platinum alkyls, only a preliminary communication has appeared with no spectra or evidence for the assignments given, so these must be regarded as tentative [30].

From the table it may be seen that most σ(M-Me) ionizations occur in the range 8–10 eV, though electronegative substituents in the metal or carbon atom will increase the range to above 10 eV.

F. Metal Carbenes

A number of metal carbene compounds of general formula $Cr(CO)_5C(X)Y$ have been studied [18]. As for the group VIIA pentacarbonyl derivatives, the $Cr(CO)_5$ group may be regarded as an octahedral fragment with the six d electrons occupying perturbed t_{2g} orbitals that are involved in back donation to the ligands. The carbene group can function as a ligand by σ donation from a σ^* M.O., which has a varying degree of carbon sp^2 character, and by accepting charge from the metal into a vacant π^* level, which is essentially a vacant carbon 2p orbital. The geometry of these molecules is such that the plane of the carbene ligand lies between the planes of the cis-carbonyl groups, so the overall molecular symmetry is C_s. I.E. data is given in Table 12.

The P.E. spectra of these compounds all show a broad band lying between 7 and 8 eV that shows no clearly defined structure, but may be curve fitted with two or three Gaussians. This band is assigned to the metal d ionizations. The first ionization energy is always significantly lower than that of $Cr(CO)_6$ (8.40 eV), demonstrating that the carbene ligand is a poorer π-acceptor than CO. Although the π^* orbital of the carbene ligand has a more favourable energy placement for interaction with the

Table 12. Ionization energies (eV) of metal carbene compounds $Cr(CO)_5C(X)Y$ [18]

X	Y	d-orbitals[a]	σ^*	ligand bands
OCH_3	CH_3	7.47, 7.89	9.89	
SCH_3	CH_3	7.35, 7.59, 7.79	9.91	11.23, 10.23
$N(CH_3)_2$	CH_3	7.12, 7.35, 7.61	9.72	10.67
NH_2	CH_3	7.45, 7.80	10.31	
OCH_3	C_4H_3O	7.37, 7.75	9.92	9.14, 10.51
NH_2	C_4H_3O	7.22, 7.52	10.30	9.21, 10.68
OCH_3	C_6H_5	7.39, 7.78	9.26	9.66, 10.06
NH_2	C_6H_5	7.25, 7.52, 7.73	9.80	9.23, 9.52, 10.50
$N(CH_3)_2$	C_6H_5	7.02, 7.26, 7.54	9.49	8.87, 10.58, 10.16, 10.96

[a] These values result from fitting the observed broad d band with the minimum numbers of Gaussians

metal d-orbitals than do the 2π CO orbitals, its geometrical situation is less favourable as a result of mixing with the substituent orbitals; and it is also singly degenerate whereas CO possesses a doubly degenerate pair of orbitals for π-acceptance.

The spectra then show a variety of bands before the CO edge at ca. 13 eV; they depend on the nature of the carbene. These are assigned by correcting orbital energies from non-parameterized M.O. calculations. The correction factors are found by comparing calculated orbital energies with experimental I.E. for related molecules where assignment is definite. The standard molecules were $Cr(CO)_6$, C_6H_6, C_4H_4O, $Cr(CO)_5C(OCH_3)CH_3$ and $Cr(CO)_5C(NMe_2)CH_3$. In all cases a band could be assigned to ionization of the metal carbene σ bond.

G. Metal Olefin Compounds

The metal-olefin bond has been studied by P.E.S. in three main classes of compounds. These are $Fe(CO)_4L$, β-diketonateML_2 and $(\eta\text{-}C_5H_5)_2ML$. In all cases the analogues where L = CO have been used for comparative assignment of spectra. Assignments as arrived at by the various authours are given in Table 13, and some representative spectra shown in Fig. 17. The results are generally discussed in terms of the commonly accepted description of the metal-olefin bond, which consists of ligand-to-metal electron donation from the filled π-orbital of the olefin into an empty metal d-orbital and metal-to-ligand back donation from a filled metal d orbital into the empty π^* orbital (the Chatt-Dewar-Duncanson model).

I. Fe(CO)₄L

The complexes $Fe(CO)_4L$ are trigonal bipyramidal with the π-bonded olefin occupying an equatorial position. The "parent" compound, $Fe(CO)_5$, shows two d-ionization bands corresponding to ionization from the $e''(3d_{xz}, 3d_{yz})$ level (9.9 eV) and the $e'(3d_{xy}, 3d_{x^2-y^2})$ level (8.6 eV)[110,111]. Substituting an olefin for one of the equatorial CO ligands lowers the symmetry from D_{3h} to C_{2v}, and the d orbitals now transform as $a_2(3d_{yz})$, $b_1(3d_{xz})$, $b_2(3d_{xy})$ and $a_1(3d_{x^2-y^2})$. The olefin π-orbital also transforms as a_1. If the olefin is unsymmetrical, the symmetry of the complex will in fact be lower, but the symmetry analysis given here still forms a useful basis for assignment.

For the purposes of assignment, the spectra are divided into three regions; the 8.4–9.5 eV region where metal d-band ionizations are expected, the 9.6–14 eV region, which contains olefinic ionizations, and the third region above 14 eV, which contains CO and further olefin ionizations. Assignments of the two lower I.E. regions are given in Table 13 for a wide variety of olefins, and some representative spectra are shown in Fig. 17. The basis for the assignments are He(I)/He(II) intensity differences and comparison with calculations[111,112]. In the case of $Fe(CO)_4C_2H_4$, discrete variations $X\alpha$ calculations have been used. In early reports[111] an incorrect

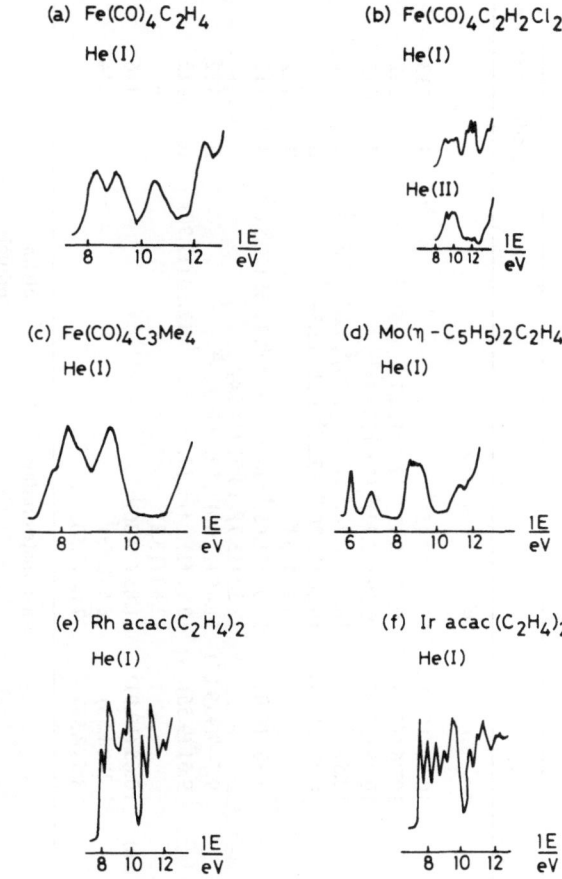

(a) $Fe(CO)_4C_2H_4$

He (I)

(b) $Fe(CO)_4C_2H_2Cl_2$

He (I)

He (II)

(c) $Fe(CO)_4C_3Me_4$

He (I)

(d) $Mo(\eta - C_5H_5)_2C_2H_4$

He (I)

(e) Rh acac$(C_2H_4)_2$

He (I)

(f) Ir acac $(C_2H_4)_2$

He (I)

Fig. 17a–f. He(I) Photoelectron spectra of some metal olefin complexes: a $Fe(CO)_4C_2H_4$; b $Fe(CO)_4$ 1,1-dichloroethylene; c $Fe(CO)_4$ tetramethylallene; d $Mo(\eta-C_5H_5)_2C_2H_4$; e Rhacac$(C_2H_4)_2$; f Iracac$(C_2H_4)_2$

spectrum of $Fe(CO)_4C_2H_4$ was given [113]: severe decomposition occured in the inlet system and a mixture of the compound, together with $Fe(CO)_5$ and C_2H_4 was recorded. A clean spectrum has now been published [113] that shows two bands at 8.38 and 9.23 eV, assigned to ionization of $d_{xy}, d_{x^2-y^2}$ and d_{yz}, d_{xz} respectively. The π-olefinic ionization is found at 10.56 eV: as the corresponding ionization of free ethylene is at 10.51 eV, very little shift is shown on complexation. This is surprising in view of the large difference in the C=C distance of free C_2H_4 (1.335 Å) and coordinated C_2H_4 (1.462 Å).

Spectra for a wide variety of olefinic ligands have been assigned with the aid of CNDO calculations [112]. These are clearly inferior to both HFS-Xα and *ab-initio* calculations giving Koopmans' theorem estimates of the C=C π-ionization, which are around 5 eV too high. It is argued, however, that they provide a useful basis for assignment in that the differences of I.E. between free and complexed olefins are well reproduced, the assignment of these latter being based in many cases on intensity variations.

Table 13. Ionization energies of metal olefin complexes[a,b,c]

$Fe(CO)_4L$					He-II	Reference
L	$d_{xy}, d_{x^2-y^2}$	d_{yz}, d_{xz}	$\pi(C=C)$	Substituent orbitals		
C_2H_4	8.38	9.23	10.56 (10.51)	12.48 (12.45)[d]	II	113
$CH_2=CH \cdot CHO$	8.69	9.42 (sh)	10.76 (10.94)	9.67 (10.11)[e] 12.9 (13.67)[d]	II	112
$CH_3CH_2=CH \cdot CHO$	8.60	9.36 (sh)	10.35 (10.38)	9.59 (9.73)[e] 12.6 (13.06)[d]	II	112
$CH_2=CH \cdot COOH$	8.66	9.36	10.57 (10.95)	10.29 (10.78)[e] 11.66 (12.00)[f], 12.9 (13.54)[d]	II	112
$CH_2=CH \cdot COOMe$	8.50	9.28	10.80 (10.74)	10.50 (11.20)[e], 12.55[f], 12.9 (13.39)[d]	II	112
$COOMe \cdot CH=CH \cdot COOMe$	8.68	9.31		10.1–11.0[g]	II	112
$CH_2=CCl_2$	8.82	9.51	9.98 (10.00)	11.05 (11.65)[h], 11.56 (12.14)[h], 12.65 (12.54)[h] 13.28 (13.7)[i], 13.91 (14.24)[h]	II	112
$trans-CHCl=CHCl$	8.72	9.49 (sh)	9.7–9.9 (9.80)	11.45 (11.90)[h], 12.0 (12.61)[h]	II	112
$trans-CHBr=CHBr$	8.74	9.45 (sh)	9.61 (9.55)	10.71 (11.04)[h], 11.28 (11.57)[h], 12.30 (12.90)[i] 12.63 (13.3)[h]	II	112
$CH_2=CHCN$	8.90, 9.05	9.70, 9.90 (sh)	10.65 (10.91)	11.85 (12.36)[j]		116
$Me_2C=C=CMe_2$	7.84	8.24	9.28 (8.53)	8.5 (8.53)[k]		114
	8.24	9.28	8.5 (8.53)	7.84 (8.53)		

β-diketonate ML_2[l]	metal d-orbitals				diketonate orbitals			olefin orbitals	He-II	Reference
	$6b_1$	$14a_1$	$5a_2$	$13a_1$	$5b_1(\pi_3)$	$10b_2(n_-)$	$12a_1(n_+)$	$9b_2$		
tmh $Ir(C_2H_4)_2$	7.32	7.76	8.17	8.79	9.12			10.3 (10.51)	II	115
acac $Ir(C_2H_4)_2$	7.36	7.83	8.37	8.86	9.35	9.51	11.24	10.41 (10.51)	II	115
tfa $Ir(C_2H_4)_2$	7.70	8.17	8.80	9.25	9.79	10.05	11.54	10.90 (10.51)	II	115
acac $Ir(C_3H_6)_2$	7.15	7.60	8.15	8.59	9.05	9.29	10.56	10.01	II	115
tfa $Ir(C_3H_6)_2$	7.57	7.99	8.61	8.99	9.54	9.71	11.01	10.36	II	115
tmh $Rh(C_2H_4)_2$	7.50		7.94		8.93		10.5	10.01 (10.51)	II	115
acac $Rh(C_2H_4)_2$	7.54		8.11		8.94	9.33	10.76	10.22 (10.51)	II	115
tfa $Rh(C_2H_4)_2$	7.96	8.56	8.7	8.8	9.44	9.79	11.21	10.64 (10.51)	II	115
hfa $Rh(C_2H_4)_2$	8.34		9.06		9.94	10.32	11.78	11.07 (10.51)	II	115
tmh $Rh(C_3H_6)_2$	7.27		7.76			8.70	10.18	9.68 (9.82)	II	115
acac $Rh(C_3H_6)_2$	7.43		7.92		8.8	9.13	10.51	9.85 (9.82)	II	115

$(\eta\text{-}C_5H_5)_2ML^m$	metal d-orbitals		C_5H_5 π-orbitals	olefin π-orbital	
	$3a_1$	$2b_1$	$2a_1 + 1b_2 + 1b_1 + 1a_2$	$1a_1$	
$Cp_2MoC_2H_4$	6.0	6.9	8.8, 9.2	11.3 (10.51)	78
$Cp_2WC_2H_4$	6.0	7.1	9.0, 9.3, 9.5	11.3 (10.51)	78
$Cp_2WC_3H_6$	5.9	7.0	8.9, 9.5	11.0 (9.82)	78

a Only assignments of the metal d-orbitals, some of the olefin bands and selected ligand ionizations are given
b Symmetry annotations are for the corresponding orbitals in the free ligands
c The values in parentheses are the ionization energies for the free ligands
d σ(C–H); e n_O(a'); f n_O(a''); g unresolved bands; h n_{Cl}; i σ(C–Cl); j π(C=N); k unperturbed π(C≡C);
l Symmetry assignments and numberings are taken from CNDO calculations on acac Co(C2H4)2
m Symmetry assignments and numbering as in Ref. 115

If this wide variety of olefin complexes is examined a general pattern emerges. The d-bands are little changed from those of $Fe(CO)_5$; two bands are always detected, but their ionization energies are slightly lower than for $Fe(CO)_5$, indicating a lower acceptor capacity for the olefins than for CO. Olefin substituent orbitals show a drop in ionization energy on complexation indicating an appreciable flow of charge from the iron to the olefin. It is concluded that π-backbonding is the most important bonding factor in these complexes. The pattern of $\pi(C=C)$ I.E. shifts varies from zero for low lying $\pi(C=C)$ orbitals to slight destabilizing shifts for higher lying $\pi(C=C)$ orbitals. They show less shift, on average, than other olefin ionizations. This may be simply interpreted as a result of two opposing effects, the stabilizing effect of donation of the π-electrons to the metal and the destabilizing effect of the accummulation of net negative charge on the ligand.

The interpretation of the spectrum of tetramethylallene iron tetracarbonyl is at variance with this general trend [114]. The assignment of Hill et al. is based on the assumption that one of the π-levels of $Me_2C=C=CMe_2$ will be perturbed by complexation and the other not. A shoulder at 8.5 eV is therefore assigned to ionization of the uncomplexed π-electrons and a band at 9.28 eV to the complexed π-electrons. The two remaining features at 7.84 and 8.24 eV are then assigned to the d-ionizations. An assignment more consistent with the pattern found by van Dam and Oskam [112] is given in Table 13 together with the original assignment. He(II) studies of this molecule would assist in deciding the correct assignment.

II. β-Diketonate ML_2

The structure of these complexes is based on a square plane with the coordinated olefins perpendicular to the molecular plane. Assignments are complicated by the fact that the low energy region contains ionizations from the four metal d-orbitals, three low lying β-diketonate orbitals and the two olefin π-orbitals. Assignments are made by comparison with the spectra of the carbonyl complexes β-diketonateM(CO)$_2$ where the olefin ionizations are absent, by consideration of substituent effects, by observation of He(I) and He(II) intensities and by using some results of extended CNDO calculations on analagous cobalt complexes. In the majority of cases reasonable detailed assignments are possible [115]. If the molecules have C_{2v} symmetry, the olefin π-orbitals transform as a_1 and b_2. The CNDO calculations suggest that the a_1 combination is highly delocalized over the molecule, while the b_2 orbital is much more localized on the olefin. As a consequence olefin bands in the low energy regions are assigned to the b_2 combination, while the a_1 orbitals are assumed to lie under the broad ligand bands in the 12−16 eV region. It is argued that destabilization of the b_2 orbital with respect to the free olefin implies a charge flow from the metal to the olefin and is due to π-backbonding being the most important factor; conversely stabilization implies σ-donation is dominant. In a number of the Rh and Ir ethylene complexes π-backbonding dominates, while in the complexes with tfa and hfa or propylene σ-donation becomes important.

III. $M(\eta\text{-}C_5H_5)_2L$

Another class of compound which has afforded study of the metal olefin bond is the bis-η-cyclopentadienyl olefin system[78]. Related compounds are also discussed in Sect. D, where their structure and an orbital scheme is described. Again the carbonyl analogues (see Sect. D.I) provide useful references for identifying ionizations from the two d-orbitals and the $e_1(\pi)$ orbitals of the cyclopentadienyl rings. In the ethylene complexes $M(\eta\text{-}C_5H_5)_2(\eta\text{-}C_2H_4)$, where M = Mo and W, extra bands are observed at 11.3 and 11.4 eV respectively. These are assigned to the olefin π-ionization, representing significant stabilization over the I.E. for free ethylene. In the tungsten propene analogue a similar band shows at 11.0 eV that is similarly assigned. Here an even greater stabilization is shown as free propylene has a π-ionization energy of 9.82 eV.

In these compounds the effect of the olefin on the metal d-ionization can also be observed. There are two d-ionization bands observed separated by 0.9 eV for Mo and 1.1 eV for W. The higher of the two bands is broader than the lower. The first band is assigned to the $3a_1$ orbital which is non-bonding, and the second to the $2b_1$ orbital which is of correct symmetry to back donate to the olefin; the bonding nature of the orbital is confirmed by the band width. The d^4 compound $Re(\eta\text{-}C_5H_5)_2H$, where no backbonding can occur, shows two bands of similar width of separation 0.6 eV (see Fig. 10). Thus evidence is provided for both components of synergic bonding; in this case the σ-donation appears to outweigh the back donation.

H. Transition Metal Diene Compounds

In this section we give consideration to the P.E. spectra of complexes where conjugated dienes, non-conjugated dienes and other four electron ligands are bonded to transition metals.

The systems studied fall into three classes. The $Fe(CO)_3$ moiety is well known for forming many stable compounds with dienes; it also forms complexes with the unstable radicals cyclobutadiene and trimethylenemethane. Few binary metal diene systems are known, but those of $Mo(C_4H_6)_3$ and $W(C_4H_6)_3$ have been studied. The third class is the group of compounds M(cp)L, where M = Co, Rh or Ir, cp = cyclopentadienyl group and L = conjugated or non conjugated diene.

The π-M.O. of butadiene are shown in Fig. 18. As the symmetry of the complexes in which butadiene is bound varies and is often very low, we will refer to the π-orbitals as π_1, π_2, π_3 and π_4. The two lowest orbitals π_1 and π_2 in free butadiene are occupied, and formally act as donor orbitals on complexing. The lowest unoccupied molecular orbital, L.U.M.O., π_3, is generally accepted to be the important orbital for back donation. The large majority of butadiene complexes show the distances C_1-C_2 and C_3-C_4 to be longer and C_2-C_3 to be shorter in the bound ligand than in the free molecules. This is consistent with occupancy of π_3 in the complex.

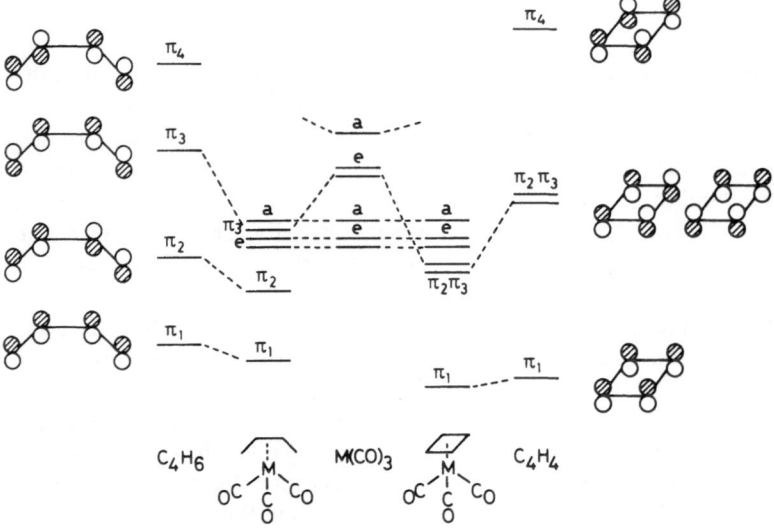

Fig. 18. Hückel M.O. for butadiene and cyclobutadiene, and interaction diagrams for $Fe(CO)_3C_4H_6$ and $Fe(CO)_3C_4H_4$

The radicals C_4H_4 and $C(CH_2)_3$ also possess four πMOs occupied by four electrons, but in these cases π_2 and π_3 are degenerate. The diagrams in Fig. 18 indicate the relative energies of C_4H_4 in a simple Hückel M.O. scheme. Those of $C(CH_2)_3$ are similar.

For a non-conjugated diene, to a first approximation of no interaction between the ene groups, degeneracies occur between π_1 and π_2 and between π_3 and π_4: these will be lifted if the two groups interact.

It is interesting to examine the effect of these different patterns of ligand π-M.O. on the orbital structure of the complexes [103].

I. Iron and Ruthenium Tricarbonyl Dienes

The iron tricarbonyl group can be treated as an octahedral fragment [103]. This suggests that its bonding capabilities reside in three frontier orbitals, of a and e symmetry, which are isolobal with those of the C—H group. In the isolated fragment these will be occupied by two electrons; the other six metal electrons occupy a perturbed "t_{2g}" set of orbitals. The $Fe(CO)_3$ group may therefore accept two electron pairs from a diene and also has the potential for transferring its own high energy pair to an unfilled diene orbital. A schematic M.O. diagram is shown in Fig. 18; π_1 interacts primarily with the a orbital and π_2 and π_3 with the e orbitals.

Lloyd et al. [16] have obtained both He(I) and He(II) spectra on $Fe(CO)_3C_4H_6$ (see Fig. 19 and Table 14). The first band is broad and complex and is assigned to four ionizations (the three d orbitals and the M.O. correlating with the diene π_3 orbital).

The second and third bands show a relative decrease in intensity in the He(II) spectrum and are assigned to the M.O. correlating with π_2 and π_1 respectively. The π_2 band shows a smaller intensity reduction in the He(II) spectrum than the π_1 band, which is taken to indicate significant metal character for π_2. Variations in the profile of the first band also produce evidence that one of its composite ionizations possesses significant ligand character. *Ab initio* calculations indicate substantial orbital relaxation on ionization, but do not reproduce the experimental sequence of bands.

The ionization energies of free trans-C_4H_6 are 9.08 eV (π_2) and 11.34 eV (π_1). The geometry change on complexing prevents an exact comparison with the bound ligand, but both orbitals are stabilized on complexing, the I.E. of π_2 being raised by 0.74 eV and that of π_1 by 0.09 eV. This contrasts with the situation found for the $Fe(CO)_4$ olefin compounds.

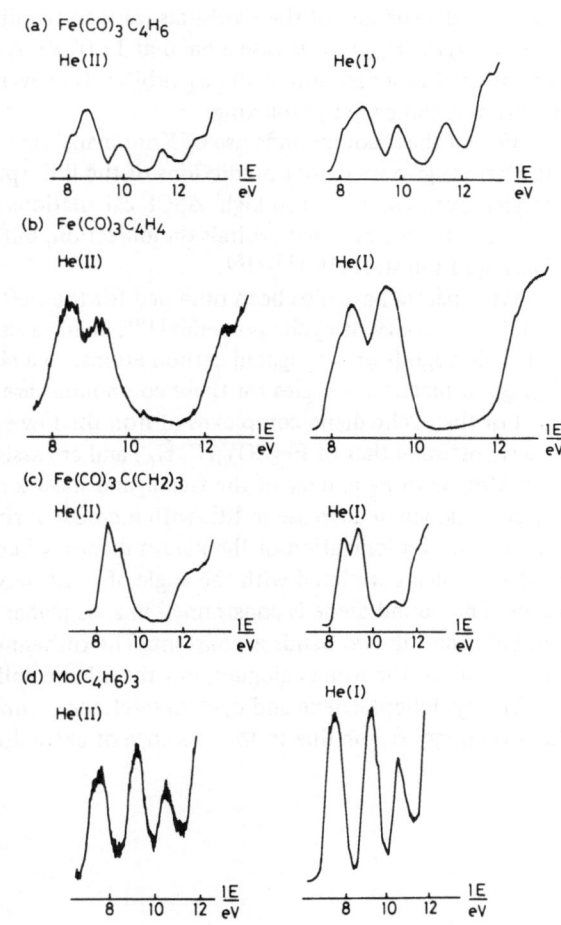

Fig. 19a–d. Photoelectron spectra of some metal diene complexes: a $Fe(CO)_3C_4H_6$; b $Fe(CO)_3C_4H_4$; c $Fe(CO)_3C(CH_2)_3$; d $Mo(C_4H_6)_3$

The He(I) and He(II) spectra of the cyclobutadiene complex, $Fe(CO)_3(C_4H_4)$, show a different band pattern in the low energy region [117]. Two bands are observed between 8 and 10 eV, the first band showing a low energy shoulder (see Fig. 19). In the He(I) spectrum the second band is the higher, but in the He(II) spectrum it drops below the first. The flat top of the second band suggests two ionization processes. These observations lead to assignment of the first band to ionization of the three metal orbitals (of t_{2g} type), and the second band to the M.O. correlating with π_2 and π_3 in the free cyclobutadiene. The different appearance from the spectrum of the butadiene complex arises from the near degeneracy of these two π orbitals. In butadiene π_2 and π_3 are well separated and π_3 is unoccupied in the ground state.

The ionization band for the M.O. correlating with π_1 of the free ligand is presumed to lie under the main band.

The closely related compound $Fe(CO)_3(C(CH_2)_3)$ has also been studied using both He(I) and He(II) radiation [118]. The low energy region is analagous to that of the cyclobutadiene compound (see Fig. 19) and is assigned in a similar manner, the first band corresponding to ionization of metal d-electrons and the second to ionization from the degenerate pair of orbitals correlating with π_2 and π_3 of the free ligand. Again the degeneracy of these orbitals leads to the differing pattern from that found for $Fe(CO)_3C_4H_6$. In this case a band at 11.07 eV is observed in the He(I) spectrum and assigned to ionization of the π_1 orbital. However its absence in the He(II) spectrum is somewhat perplexing.

For all these compounds use of Koopmans' theorem and *ab initio* SCF M.O. calculations give very poor predictions of the P.E. spectrum as the metal ionization energies given are much too high. ΔSCF calculations indicate considerable orbital relaxation for metal based orbitals on ionization, but still do not achieve the correct ordering of ion states [16,117,118].

P.E. spectra have also been obtained for the $Fe(CO)_3$ and $Ru(CO)_3$ moieties bound to a variety of cyclic polyenes [119]; in all cases binding of the ring to metal occurs through four conjugated carbon atoms, so a close analogy exists with $Fe(CO)_3$-(C_4H_6). Ionization energies for these compounds are given in Table 14.

For the cyclic diene complexes of iron the low energy regions of the P.E. spectra closely resemble that of $Fe(CO)_3(C_4H_6)$ and are assigned accordingly. The bands correlating with π_2 and π_1 of the free ligand show a regular variation of I.E. with ring size, namely a decrease in I.E. with increase in ring size. This is in contrast to the pattern of π ionization of the parent dienes, where the separation of the two bands has been correlated with the angle of twist between the conjugated double bonds. The bound diene is constrained in a cis planar conformation, so the separation between the two bands is constant. The ruthenium compounds have similar P.E. spectra to the iron analogues, but the d-band splitting is greater.

The cycloheptatriene and cyclooctatetraene complexes show additional bands in the low energy region due to the presence of extra double bonds in the ring.

II. Tris-Butadiene Metal Complexes

Tris-butadiene compounds of molybdenum and tungsten have been synthesized. Both He(I) and He(II) P.E. spectra have been obtained[120] (see Table 14 and Fig. 19). The low energy regions show three bands that, surprisingly, vary very little in relative intensity with photon energy. The clear implication of this is that there is no d-band and all the high lying M.O. have approximately the same amount of ligand character: the bands are therefore correlated with π_3, π_2 and π_1 of butadiene. Such substantial occupancy of π_3 indicates a large transfer of charge to the butadiene ligands. This is supported by the π_1 ionization energy of 10.5 eV, which lies well below that of butadiene (11.34 eV). At variance with this interpretation is the reported crystal structure of $Mo(\eta-C_4H_6)_3$, which shows alternation of C–C distances in the bound ligand and bond lengths very close to those of butadiene itself. Occupancy of π_3 would be expected to equalize the C–C bond lengths, which is the pattern found in all other butadiene complexes.

III. Metal Cyclopentadienyldiene Complexes

The 18 electron rule constrains this class of compound to the cobalt group and P.E. spectra have been obtained for derivatives of Co, Rh and Ir [121]. A M.O. scheme for the $M(C_5H_5)$ fragment is given in Fig. 20, the resemblance to that for a bis-cyclopentadienyl molecule being evident. For the metals of the cobalt group, two electrons occupy the antibonding e_1^* orbital: the second and third row metals are expected to have a more covalent interaction with the e_1 orbitals of the cyclopentadienyl rings than the first row metal, with the consequence that this antibonding level may be higher in energy for Rh and Ir than for Co. The e_1 orbitals interact with π_2 and π_3 of a conjugated diene ligand or with the π and π^* orbitals of a non-conjugated diene. Possible interaction schemes are shown in Fig. 20.

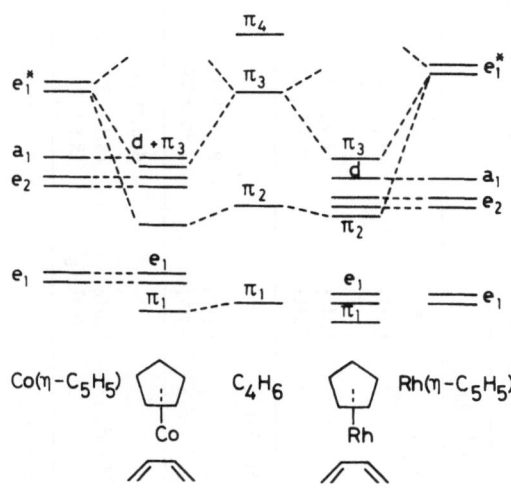

Fig. 20. Interaction diagrams for $M(\eta-C_5H_5)(\eta-C_4H_6)$ where M = Co and Rh

95

Table 14. Ionization energies of metal diene and related compounds[a]

LM(CO)₃					He(II)	References
M = Fe	L =	d-electrons	1a₂(π₂)	1b₁(π₁)	σ bands	
	C₄H₆	(8.23), 8.82	9.93	11.52	12.94	II	16, 122, 123
	C₆H₈	(7.98), 8.56	9.33	11.04	12.17		119
	C₇H₁₀	(7.78), 8.46	9.12	10.86	11.71		119
	C₈H₁₂	(7.45), 8.27	8.87	10.44	10.87		119
M = Ru	L =	d-electrons	1a₂(π₂)	1b₁(π₁)	σ bands	
	C₆H₈	8.01, 8.91	9.39 (sh)	11.01	11.83		119
	C₇H₁₀	7.96, 8.94	9.40 (sh)	10.84	11.64		119
M = Fe		d-electrons	π₂, π₃	π₁	σ	
	C₄H₄	(8.17), 8.45	9.21	12.81	13.69	II	117, 122, 123
	C(CH₂)₃	8.62	9.26	11.07	12.11	II	118, 124
M = Fe		d-electrons	ligand π		ligand σ	
	C₇H₈	(7.76), 8.39	8.78, 10.23, 11.10		11.82		119
	C₈H₈	7.84	8.74, 10.61, 11.63				119

M(C₄H₆)₃					He(II)	References
M =	d + (π₃)	1a₂ (π₂)	1b₁(π₁)	σ		
Mo	(7.23), 7.44	9.10	10.49	12.25	II	120
W	(7.34), 7.74	9.18	10.48	12.30	II	120

cp ML complexes	π₃	d + π₂	e₁(πcp)	π₁	ligand bands	He(II)	References
Rh(η-C₅H₅) (η-C₄H₆)	7.26	8.32, 8.81,	10.17	10.99	13.05	II	121
Ir(η-C₅H₅) (η-C₄H₆)	7.13	8.02, 8.29, 8.71,	10.52	10.93	12.85	II	121
Ir(η-C₅H₅) (η-MeC₄H₅)[b]	7.21	8.00, 8.29, 8.76, 9.08	10.50	–	12.73	II	121
Rh(η-C₅H₅) (η-C₆H₈)[c]	7.21	8.11, 8.54	9.87	10.40	11.47	II	121
Rh(η-C₅H₅) (η-MeC₄H₄COCH₃)[d]	7.40	8.27, 8.68	10.17	10.99	12.93, 13.93	II	121
Ir(η-C₅H₅) (η-MeC₄H₄COCH₃)[d]	7.15	7.94, 8.24, 8.65, 9.06	10.28	10.72	11.64, 12.59, 13.98	II	121
Rh(η-C₅H₅) (η-MeC₄H₄CHO)[e]	7.41	8.42, 8.75	10.25	11.01	12.68, 13.74	II	121

	d	π_2	$e_1(\pi cp)$	π_1			
Co$(\eta\text{-}C_5H_5)(\eta\text{-}C_6H_8)$[c]	6.92, 7.43, 7.97	8.47	9.27	10.43	11.21, 12.28, 13.51	II	121
	π_3	d		$e_1(cp\pi) + \pi_1$			
			π_2	π_1			
Rh$(\eta\text{-}C_5Me_5)(\eta\text{-}C_2H_4)_2$	7.12	8.27, 8.97	9.57	10.25	11.56, 12.60, 13.70	II	121
Rh$(\eta\text{-}C_5H_5)(\eta\text{-}C_8H_{12})$[f]	7.07	7.89, 8.12, 8.57, 8.84	9.27	9.82	10.70, 11.08, 12.27, 12.99, 13.53	II	121
Rh$(\eta\text{-}C_5Me_5)(\eta\text{-}C_8H_{12})$[f]	6.36	7.67, 8.16	9.11		10.23, 11.42	II	121
Rh$(\eta\text{-}C_5H_5)(\eta\text{-}C_8H_8)$[g]	7.07	7.91, 8.65, 9.00	9.93	10.77	11.31, 12.29, 13.90	II	121
	d + π_3		$\pi_2\ \pi_1\ e_1$				
Co$(\eta\text{-}C_5H_5)(\eta\text{-}C_8H_{12})$[f]	6.73,	7.25	8.58	9.01	10.43, 10.71, 11.42, 12.24, 13.55,	II	121
Co$(\eta\text{-}C_5Me_4Et)(\eta\text{-}C_8H_{12})$[f]	6.69	7.21	8.13	8.84	11.37, 11.93, 13.41	II	121

[a] This table includes complexes of conjugated dienes, non-conjugated dienes, cyclobutadiene and trimethylenemethyl

b CH_3 ... η^4-2-methylbuta-1,3-diene

c η^4-cyclohexa-1,3-diene

d CH_3—...—C(CH$_3$)=O η^4-hepta-3,5-diene-2-one

e CH_3—...—C(H)=O η^4-hexa-2,4-dien-1-al

f η^4-cycloocta-1,5-diene

g η^4-cycloocta-1,3,5,7-tetraene

J.C. Green

Some representative spectra are shown in Fig. 21, and ionization energies together with band assignments are given in Table 13.

The spectra of $Rh(\eta\text{-}C_5H_5)(\eta\text{-}C_4H_6)$ and $Ir(\eta\text{-}C_5H_5)(\eta\text{-}C_4H_6)$ are clearly related, though in the Ir case more structure is seen in the second complex band. From intensity changes this band is clearly predominantly metal in character and is assigned to the e_2 and a_1 d orbitals (where the symmetry labels refer to their $M(\eta\text{-}C_5H_5)$ fragment origin) together with π_2, the M.O. resulting from interaction between π_2 of the diene and one of the fragment e_1 orbitals. The first band, which drops substantially in relative intensity between the He(I) and He(II) is assigned to the M.O. resulting from interaction of the diene π_3 orbital with the other e_1 fragment orbital. This orbital appears to have a high degree of ligand character, which is very unusual for the first ionization of an organometallic with d electrons.

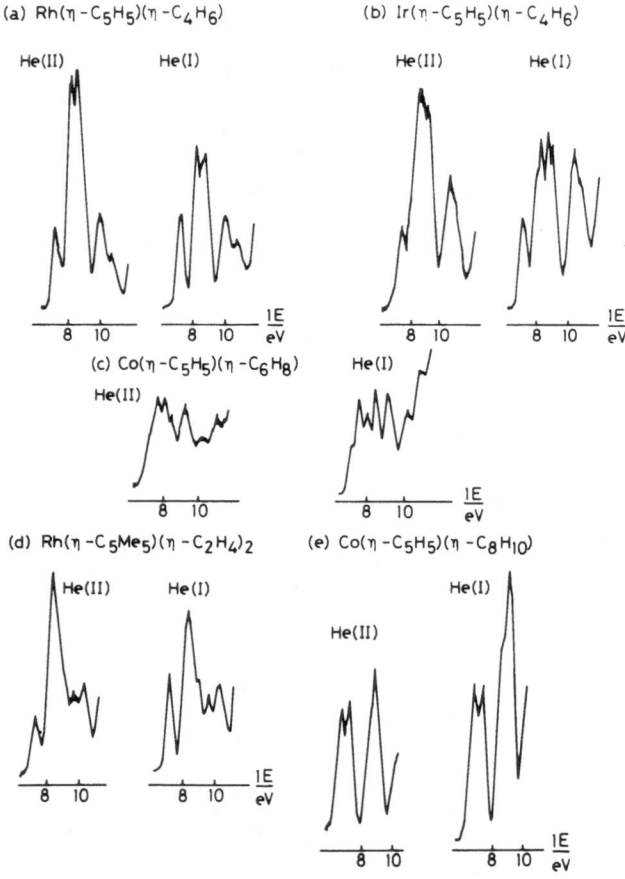

Fig. 21a–e. Photoelectron spectra of some metal cyclopentadienyldiene complexes:
a $Rh(\eta\text{-}C_5H_5)(\eta\text{-}C_4H_6)$; b $Ir(\eta\text{-}C_5H_5)(\eta\text{-}C_4H_6)$; c $Co(\eta\text{-}C_5H_5)(\eta\text{-}C_6H_8)$; d $Rh(\eta\text{-}C_5Me_5)$-$(\eta\text{-}C_2H_4)_2$; e $Co(\eta\text{-}C_5H_5)(\eta\text{-}C_8H_{10})$

The third band (10.17 eV) is assigned to the C_5H_5 e_1 ionization and the fourth band (10.99 eV), which is less intense than the third, to the singly degenerate π_1 of the butadiene ligand. The π_1 ionization at 10.99 eV is about 0.4 eV lower than the π_1 ionization in the free ligand, indicating a transfer of charge to the ligand.

Other rhodium and iridium derivatives show spectra with a similar band pattern, with the iridium compounds having the broader and more structured d-band in line with greater ligand field splittings for the heavier metal.

A direct comparison may be made between Co and Rh binding to conjugated dienes in the compounds $M(\eta\text{-}C_5H_5)(\eta\text{-}C_6H_8)$ where the metal is bound to cyclohexa-1,3-diene. The rhodium spectrum closely resembles that of $Rh(\eta\text{-}C_5H_5)$-$(\eta\text{-}C_4H_6)$ whereas the cobalt spectrum is rather different. The latter bears a greater resemblance to the spectra found for diene $Fe(CO)_3$ complexes: the lowest energy band is a complex d-band presumably containing the π_3 ionization, whereas the π_2 ionization band appears above the d-band just separated from it. The cp e_1 ionization and the diene π_1 ionization occur in the same order as in the Rh compound, but at lower energies. All the differences between the Co and Rh and Ir compounds can be attributed to lower energies of the Rh and Ir d-orbitals.

The compound $Rh(\eta\text{-}C_5Me_5)(\eta\text{-}C_2H_4)_2$ may be taken as an example of rhodium binding to a non-conjugated diene. The first ionization band shows a high degree of ligand character and is assigned to π_3; the second band is associated with ionization of metal d-electrons (e_2 and a_1). The third band (9.57 eV) of low intensity is presumed due to a single π ionization from the ethylene ligands, most likely from the π_2 orbital. The fourth band (10.25 eV) is assigned to ionization of the e_1 orbitals of the C_5Me_5 group and may well also contain the π_1 ionization. Other rhodium non-conjugated diene compounds show a similar pattern, with bands shifting in line with the known inductive effects of substituents.

Cobalt can be compared directly with rhodium, in the form of the compounds $M(\eta\text{-}C_5H_5)(\eta\text{-}C_8H_{12})$ where the diene ligand is cycloocta-1,5-diene. The same contrast is found, as for the conjugated diene complexes, the first ionization feature for the cobalt compounds being the d-band; this band is also presumed to contain the ionization correlating with π_3. In this case the bands correlating with π_1 and π_2 overlap with that due to the cyclopentadienyl ring e_1 ionization, and they may not be distinguished.

These rather striking differences, which are consistently found, between the ionization patterns depending on the metal may well reflect a considerable difference in the nature of the HOMO between Co and Rh and Ir. If this is the case, there might well be significant chemical consequences.

In the cyclopentadienyl metal diene compounds, back donation from the metal to the diene appears equally facile irrespective of whether the diene is conjugated or not, but considerably more substantial for Rh and Ir than for Co.

I. Transition Metal Allyl Compounds

The allyl radical has three π-orbitals which are diagrammatically represented in Fig. 22. The orbital, π_2, which is singly occupied in the neutral radical, is expected to mix strongly with metal d-orbitals, the degree of mixing being very sensitive to the energy of the d-orbitals in the metal fragment. The consequences of this for the photoelectron spectra of metal allyl compounds is that the ionization energy of the resultant M.O. (which we will label π_2 for convenience indicating the organic component origin) will vary widely between complexes.

The simplest P.E. spectra to assign in this class are those where only a single allyl group is bound to the metal, so these will be considered first.

(a)

(b)

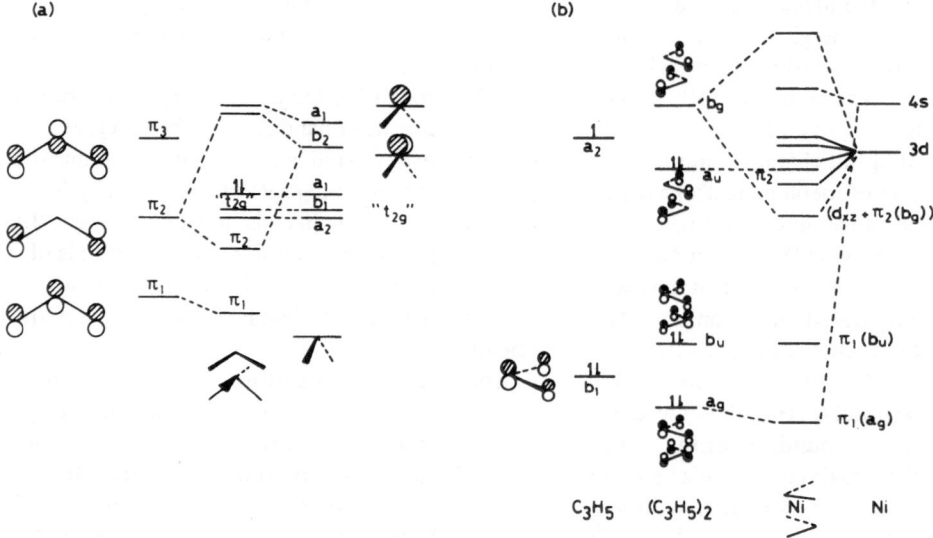

Fig. 22a, b. Interaction scheme for a Mn(CO)$_4$ (η-C$_3$H$_5$), b Ni(η-C$_3$H$_5$)$_2$ [130)]

I. $Mn(\eta\text{-}C_3H_5)(CO)_4$

Though the structure of Mn(η-C$_3$H$_5$)(CO)$_4$ is unknown, it is likely that the Mn(CO)$_4$ fragment will have C$_{2v}$ geometry as in this conformation it possesses low lying orbitals of a_1 and b_2 symmetry suitable for interacting with π_1 and π_2 of the allyl group[103)]. The proposed structure is indicated in Fig. 22, together with an interaction diagram. The P.E. spectrum (see Fig. 23)[125)] is simply assigned. The first band is assigned to ionization of the "t$_{2g}$" type electrons involved in backbonding to the CO ligands, the second band (9.16 eV) is assigned to ionization of the π_2 M.O. and the third to ionization of π_1.

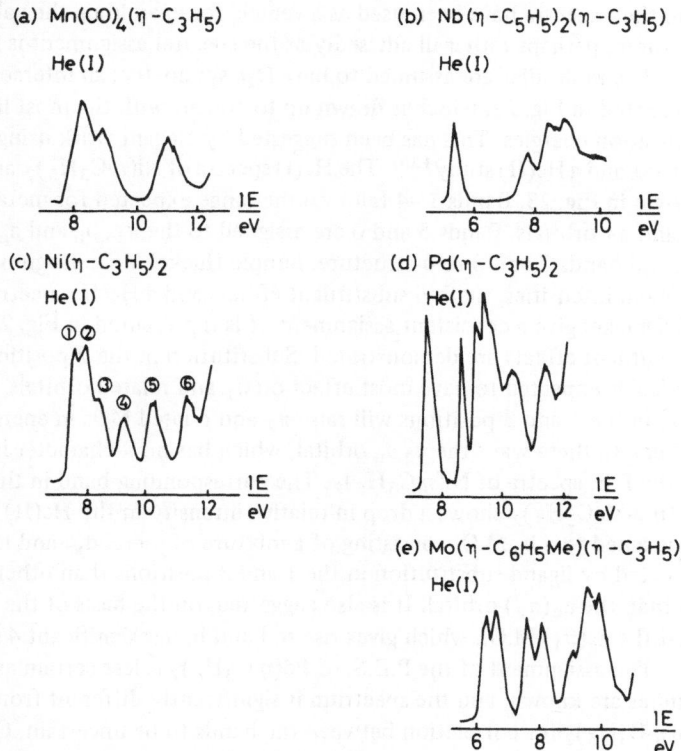

Fig. 23a–e. Photoelectron spectra of some metal allyl complexes: a Mn(CO)$_4$ (η-C$_3$H$_5$); b Nb(η-C$_5$H$_5$)$_2$ (η-C$_3$H$_5$); c Ni(η-C$_3$H$_5$)$_2$; d Pd(η-C$_3$H$_5$)$_2$; e Mo(η-C$_6$H$_5$Me) (η-C$_3$H$_5$)$_2$

II. Nb(η-C$_5$H$_5$)$_2$ (η-C$_3$H$_5$)

The bent sandwich M(η-C$_5$H$_5$)$_2$ also has orbitals of suitable symmetry for interacting with π_1 and π_2 of the allyl group[78]. The d^2 compound Nb(η-C$_5$H$_5$)$_2$-(η-C$_3$H$_5$) shows a lone pair ionization at 5.7 eV followed by a band at 8.0 eV that can clearly be assigned to ionization of π_2. In this case the π_1 ionization band has not been identified and is likely to lie under the main band resulting from orbitals of the C$_5$H$_5$ rings.

III. Bis-η-Allyl Metal Compounds

The P.E. spectra of (d^8) bis-η-allyl compounds of nickel and palladium pose severe assignment problems. The spectra display a tantalizing amount of structure; but as the d-orbitals ionize in the same region as the π_2 M.O., they are difficult to disentangle. The literature on these compounds[14,126–130] includes various M.O. calculations,

101

and their spectra have been used as a vehicle for examining the validity of Koopmans' theorem, perhaps rather ill advisedly as the spectral assignment is not clear cut.

The molecules are assumed to have C_{2h} symmetry; an interaction scheme is presented in Fig. 22, which is drawn up to concur with the most likely ordering of ionization energies. This has been suggested by elegant work using methyl substituent effects and a He(II) study [130]. The He(I) spectra of $Ni(\eta\text{-}C_3H_5)_2$ and $Pd(\eta\text{-}C_3H_5)_2$ are shown in Fig. 23. Bands 1–4 fall into the range expected for metal d-orbitals and the ligand π_2 orbitals. Bands 5 and 6 are assigned to the π_1, b_u and a_g orbitals respectively, and bands 7 and 8 to σ-structure. Simple Hückel M.O. assignments, consideration of band intensities, methyl substituent effects and a He(II) spectrum of bis(2-methyl-allyl)nickel give a consistent assignment; it is represented in Fig. 24 where the methyl substituent effects are demonstrated. Substitution in the 2 position of the allyl radical is expected to have most effect on π_1 and related orbitals, whereas substitution in the 1 and 3 positions will raise π_2 and related MOs in energy more than others. In these ways the π_2 a_u orbital, which has no d-character is assigned to band 2 in the P.E. spectra of $Ni(\eta\text{-}C_3H_5)_2$. The corresponding band in the spectrum of $Ni(\eta\text{-}2\text{-}MeC_3H_4)_2$ shows a drop in relative intensity in the He(II) spectrum. Band 4 is assigned to a b_g M.O. consisting of a mixture of metal d_{xz} and ligand π_2. It is more affected by ligand substitution in the 1 and 3 positions than other d bands but less so than the $a_u(\pi_2)$ orbital. It is also suggested, on the basis of the He(II) spectrum, that the $a_g(\pi_1)$ M.O., which gives rise to band b, has significant 4 s character.

The assignment of the P.E.S. of $Pd(\eta\text{-}C_3H_5)_2$ is less certain as no detailed studies are known, and the spectrum is significantly different from that of $Ni(\eta\text{-}C_3H_5)_2$ for correlation between the bands to be uncertain. Only the π_1, a_g and b_u ionizations may be unambiguously identified.

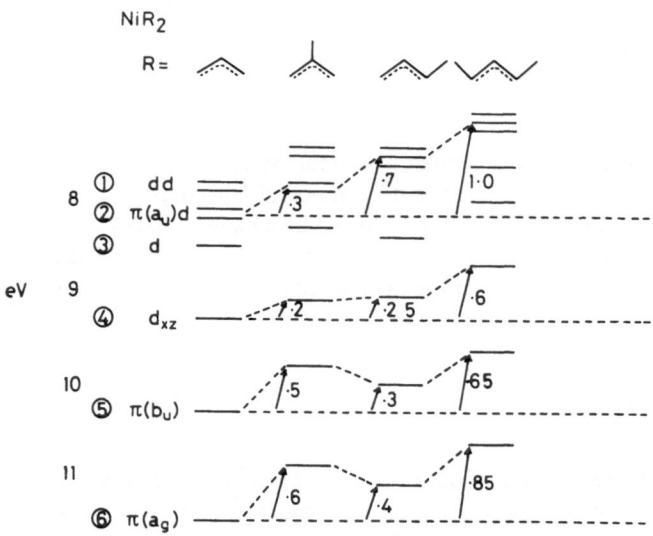

Fig. 24. Correlation diagram for the I.E. of allyl-nickel complexes adapted from reference [130]

Table 15. Ionization energies (eV) of some η^3-allyl metal compounds

			He (II)		References
	d + π₂			π₁	
Ni(η-C₃H₅)₂	7.76, 8.19, 8.58, 9.40			10.38, 11.55	129, 130
Ni(η-CH₂C(Me)CH₂)₂	7.53, 7.91, 8.32, 9.22			9.86, 10.93	130 (II)
Ni(η-MeCHCHCH₂)₂	7.53, 8.00, 8.40, 9.13			10.10, 11.15	130
Ni(η-MeCHCHCHMe)₂	7.22, 7.68, 8.10, 8.78			9.73, 10.70	130
Pd(η-C₃H₅)₂	7.59, 8.74, 9.22, 9.52, 9.73			10.43, 11.56	129
Cr(η-C₃H₅)₃	7.26, 7.95, 8.89			10.76	131
	d	π₂	e₁(π)	π₁	
Nb(η-C₅H₅)₂(η-C₃H₅)	5.7	8.0	8.6, 9.2		78
Mo(η-C₆H₅Me) (η-C₃H₅)₂ᵃ	6.44, 6.86	7.74, 8.29	9.18	9.69, 10.08	125
Mn(η-C₃H₅) (CO)₄	8.2, 8.58	9.16		11.14	125

ᵃ e₁(π) and π₁ bands overlap and may not be distinguished

The compound $Mo(\eta\text{-}C_6H_5Me)(\eta\text{-}C_3H_5)_2$ is a bis-allyl compound in which the four d electrons are clearly separated from the π_2 ionizations (see Fig. 23 and Table 15)[125].

IV. $Cr(\eta\text{-}C_3H_5)_3$

Only a rough assignment of the P.E. spectrum[131] of this compound may be made as it is another case where d orbitals and π_2-related M.O. ionize in the same spectral region. He(II) studies indicate that the shoulder at 8.89 eV arises from an M.O. with significant ligand character. A band associated with some π_1 M.O. may be clearly identified at 10.76 eV.

J. A Miscellany of "Half-Sandwich" Compounds

A half-sandwich compound is one in which a metal is π-bonded to just one carbocyclic ring. Also included in this section are related compounds in which the organic group's delocalized π-system is not cyclic, and some in which a carborane fragment is bound to the metal.

In deriving a qualitative M.O. scheme by fragment analysis, two approaches are possible. Either the interaction of the orbital structure of the half-sandwich with that of the other ligands bound to the metal may be considered or, alternatively, the interaction of the ML_m unit, where L is usually a carbonyl, with the carbocyclic ring may provide a more convenient approach. The two methods are complementary, the former being more fruitful when m is low and the latter when m is high. In Fig. 25, interaction schemes are given for $M(C_5H_5)(CO)_n$ where n = 2,3 and 4.

All compounds considered in this section conform to the 18 electron rule and are subclassified according to their d^n configuration.

I. d^4 Compounds

The only member of this class is the molecule $V(\eta\text{-}C_5H_5)(CO)_4$; its P.E. spectrum is given in Fig. 26. Two d bands are observed, which, on the level scheme given in Fig. 25, are assigned to the $b_2(d_{xy})$ and $a_1(d_{z^2})$ orbital [125]. The b_2 orbital, which is expected to be involved in back donation to the CO ligands, is assigned to the broader of the two bands, while the a_1 orbital, which is expected to be more non-bonding in character, is assigned to the sharper of the two bands overlying the broader one.

The ring e_1 orbitals are assigned to the ionization band at 9.52 eV occurring at somewhat higher energy than the ring e_1 ionizations in vanadocene (8.4–9.0 eV).

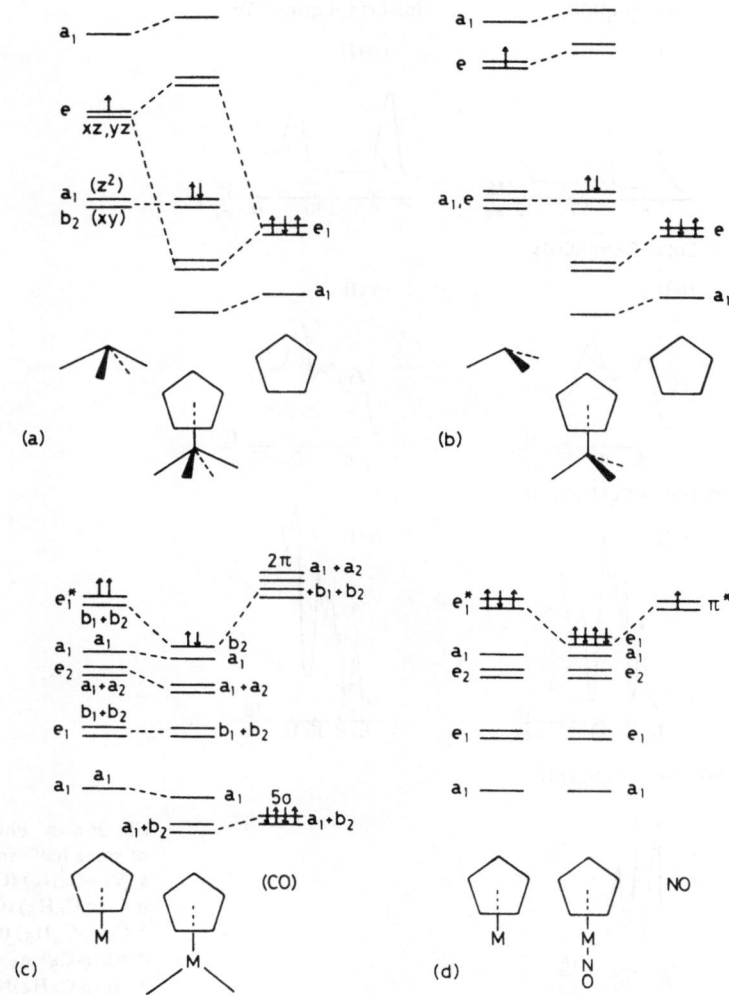

Fig. 25a–d. Interaction diagram for $M(\eta\text{-}C_5H_5)(CO)_n$ where n = 4 (a), 3 (b), 2 (c) and $M(\eta\text{-}C_5H_5)NO$ (d)

II. d^6 Compounds

Most of the compounds in this class contain an $M(CO)_3$ fragment. This has three low lying t_{2g} type orbitals, for holding the six d-electrons, and three frontier orbitals of a and e symmetry, which act as acceptors and are well matched with the a and e_1 orbitals of an unsaturated carbocyclic ring[103].

The spectrum of a typical compound, $Cr(\eta\text{-}C_6H_6)(CO)_3$, is shown in Fig. 26[20]. In this and other arene metal tricarbonyl compounds, the d-ionizations occur as one broad band, the a_1 and e bands not being resolved. Structure is observed for mesityle-

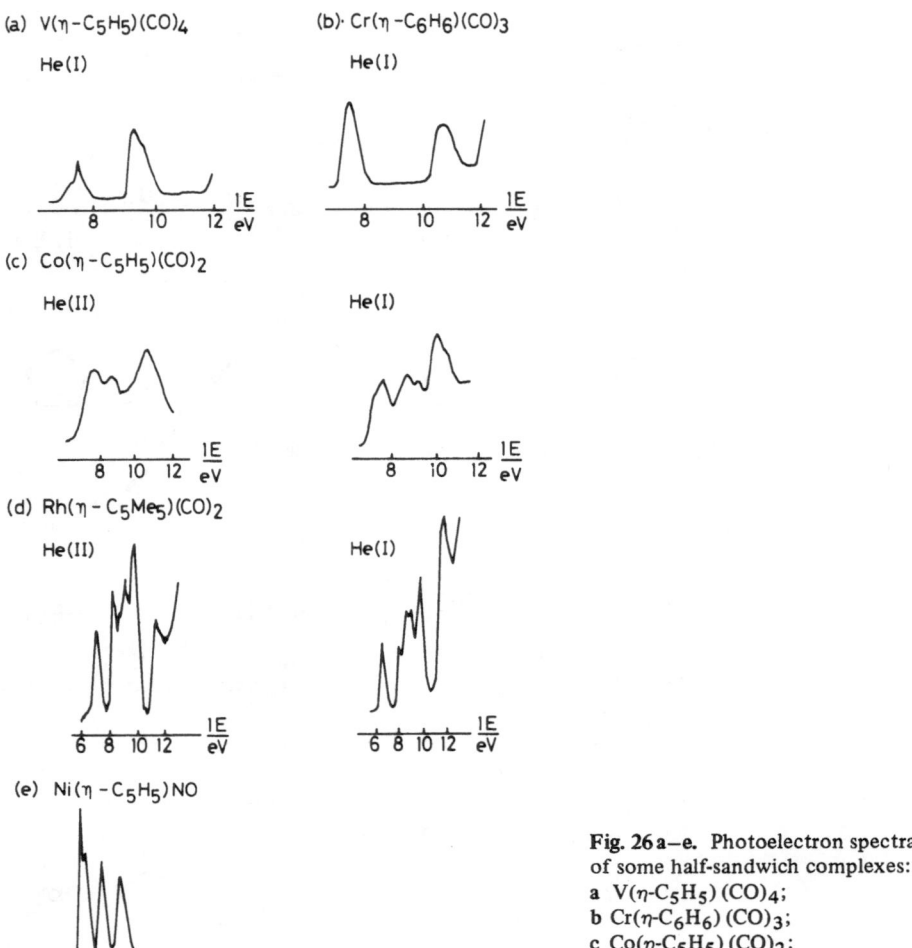

(a) V(η-C$_5$H$_5$)(CO)$_4$

He(I)

(b) Cr(η-C$_6$H$_6$)(CO)$_3$

He(I)

(c) Co(η-C$_5$H$_5$)(CO)$_2$

He(II)

He(I)

(d) Rh(η-C$_5$Me$_5$)(CO)$_2$

He(II)

He(I)

(e) Ni(η-C$_5$H$_5$)NO

Fig. 26 a–e. Photoelectron spectra of some half-sandwich complexes:
a V(η-C$_5$H$_5$)(CO)$_4$;
b Cr(η-C$_6$H$_6$)(CO)$_3$;
c Co(η-C$_5$H$_5$)(CO)$_2$;
d Rh(η-C$_5$Me$_5$)(CO)$_2$;
e Ni (η-C$_5$H$_5$)NO

ne tungsten tricarbonyl, but this is strongly reminiscent of that found on the d-band of W(CO)$_6$ and is assigned to spin-orbit splitting [132]. The I.E. of the π(e) orbital of the arene ring is raised compared to the free arene, 1.45 eV for benzene and 1.7 eV for mesitylene.

The d-band, which is unresolved for Mn(η-C$_5$H$_5$)(CO)$_3$ [133], shows structure when one of the carbonyls is substituted by N$_2$ or NH$_3$ [134]. Also the bands occur on average at lower I.E. Related calculations indicate that this is because N$_2$ is a poorer π-acceptor than CO whereas NH$_3$ has no π-acceptor capacity and is a strong σ-donor. This results in rather low I.E. for Mn(η-C$_5$H$_5$)(CO)$_2$N$_2$.

If the organic group is not cyclically conjugated, the spectra appear more complex [132] as the d-band shows structure and there is no longer degeneracy among the ligand π-orbitals. For the η^6-cycloheptatriene complexes two d-bands are observed, the splitting presumably being due to the unequal interaction of the e levels of the $M(CO)_3$ fragment with the unoccupied π-orbitals of the triene fragment. The third and fourth bands correlate with the two lowest π-orbitals of the cycloheptatriene ring; the third band occurs 0.75 eV higher than in the free ligand while the fourth band occurs 1.1 eV higher. A parallel study on dienyl compounds of $Mn(CO)_3$ shows that the principal bonding interaction is with the upper occupied level of the pentadienyl group rather than the second π-level [135].

The ferraboranes, although arguably organometallics, may be regarded as closely related [136]. In the spectra of the compounds listed in Table 16 the a and e d-ionizations associated with the $Fe(CO)_3$ fragment may be identified; it is found that $C_2B_5H_5$, $C_2B_3H_7$, B_4H_9 and B_5H_9 all increase the d-ionization energies compared with C_4H_4, the closest organic analogue. The splitting between the a and e orbitals is a function of the cluster and is larger for bipyramidal (closo) than for pyramidal (nido) structures. The ligand bands are assigned largely by analogy with the borane or carborane formed when the $Fe(CO)_3$ group is replaced by BH: the results provide strong support for the isolobic nature of these two groups.

III. d^8 Compounds

Spectra of $Co(\eta\text{-}C_5H_5)(CO)_2$ and $Rh(\eta\text{-}C_5Me_5)(CO)_2$ are shown in Fig. 26 [121]. The cobalt compound readily undergoes thermal decomposition at low temperatures and pressures, and as a consequence a clean spectrum is difficult to obtain. The pattern of low energy region bands differs for the two compounds partly as a result of the greater splittings found for second row metals (cf. the different patterns for the M(cp) dienes given in Sect. H. III) and partly due to the methyl substitution on the cyclopentadienyl ring, which lowers the energy of the e_1 cp(π) band in the rhodium compound.

The orbital structure of these compounds is most easily visualized by considering the interaction of the Mcp unit with two CO ligands. The CO ligands provide an acceptor orbital, b_1, for the two electrons occupying the e_1 antibonding orbital of the Mcp fragment.

In the spectrum of $Rh(\eta\text{-}C_5Me_5)(CO)_2$ this b_1 ionization occurs as a well separated band at low energy (6.66 eV), indicating its M-cp e_1^* antibonding character. For $Co(\eta\text{-}C_5H_5)(CO)_2$ the b_1 ionization overlaps with other d bands.

In the rhodium case, considerable back donation appears to occur from the metal $a_1 + a_2$ (e_2) levels to the CO ligands as these orbitals show a greater ionization energy than the cp e_1 orbitals. In the cobalt compound the I.E. ordering is the more conventional $e_2 < e_1$.

IV. d^{10} Compounds

In the case of $Ni(\eta\text{-}C_5H_5)NO$ and $Ni(\eta\text{-}C_5H_4Me)NO$ the orbital structure is best approached by considering the Mcp unit first and subsequently the interaction with NO (see Fig. 25). The spectra of these two compounds (see Fig. 26) were assigned

Table 16. Ionization energies (eV) of half-sandwich compounds

	d-ionizations[a]	e_1 or ligand band	References
d^4 compounds			
$V(\eta\text{-}C_5H_5)(CO)_4$	7.36 (b_2) 7.59 (a_1)	9.52, 9.76	125
d^6 compounds			
$Mn(\eta\text{-}C_5H_5)(CO)_3$	8.05, 8.40 (e + a)	9.90, 10.29	133, 135
$Re(\eta\text{-}C_5H_5)(CO)_3$	8.13, 8.52, 8.76	10.18, 10.59	133
$Cr(\eta\text{-}C_6H_6)(CO)_3$	7.42	10.70	20
$Cr(\eta\text{-}C_6H_5Me)(CO)_3$	7.24	10.41	125
$Cr(\eta\text{-}C_6H_3Me_3)(CO)_3$	7.20	10.08	132
$Mo(\eta\text{-}C_6H_3Me_3)(CO)_3$	7.35	10.08	132
$W(\eta\text{-}C_6H_3Me_3)(CO)_3$[b]	7.20, 7.45	10.17	132
$Mn(\eta\text{-}C_5H_5)(CO)_2N_2$	7.54, 7.89, 8.07	9.78, 10.17	134
$Mn(\eta\text{-}C_5H_5)(CO)_2NH_3$[c]	6.63, 6.99, 7.36	9.15	134
$Cr(\eta\text{-}C_7H_8)(CO)_3$	7.30, 7.73	9.32	132
$Mo(\eta\text{-}C_7H_8)(CO)_3$	7.46, 7.94	9.29	132
$W(\eta\text{-}C_7H_8)(CO)_3$	7.55, 8.05	9.40	132
$Mn(\eta^5\text{-}C_6H_7)(CO)_3$	8.06	8.59, 10.25	135
$Mn(\eta^5\text{-}C_7H_9)(CO)_3$	7.86, 8.10	8.67, 9.97	135
$Mn(\eta^5\text{-}C_7H_7)(CO)_3$	7.66, 7.86	8.33, 10.33	135
$Fe(C_2B_3H_5)(CO)_3$	8.6, 9.1	9.9	136
$Fe(C_2B_3H_7)(CO)_3$	(8.7), 8.9	10.6	136
$Fe(B_4H_8)(CO)_3$	(8.6), 8.9	10.3	136
$Fe(B_5H_9)(CO)_3$	(8.9), 9.2	8.4, 10.8	136
d^8 compounds			
$Co(\eta\text{-}C_5H_5)(CO)_2$	7.51 (b_1) 7.78 (a_1)		
	8.65, 9.17 (e_2)	9.90, 10.31	121
$Rh(\eta\text{-}C_5Me_5)(CO)_2$	6.66 (b_1) 8.00 (a_1)		
	9.52 (e_2)	8.51, 8.84	121
d^{10} compounds			
$Ni(\eta\text{-}C_5H_5)(NO)$	8.29 (e_1) 8.48 (a_1)		
	9.52 (e_2)	10.27	137
$Ni(\eta\text{-}C_5H_4Me)(NO)$	8.09, 8.32 ($e_1 + a_1$)		
	9.30 (e_2)	10.15	137

[a] Symmetry labels assigned to some bands are explained in the text
[b] Splitting of d band due to spin-orbit coupling
[c] Compound underwent partial decomposition on sublimation and bands were obtained by spectral stripping

with the aid of *ab initio* ΔSCF calculations, which, although not predicting the correct ordering of the ion states, gave a reasonable grouping of the ionization energies[137]. The original assignments were made largely by considering intensities and I.E. shifts on methyl substitution and have subsequently been confirmed by He(II) studies[138]. The high ligand content of the first e_1 band is a reflection of its origin as an M-cp e_1^* antibonding orbital.

K. Concluding Remarks

From this survey of the P.E. spectra of organometallic compounds, it is clear that, in the large majority of cases, certain assignment of the spectra has been achieved and the ordering of ion states established. This has enabled comparison of bonding between metals, and both between and within various ligand classes.

Some uncertainty still surrounds the interpretation of the intensity changes with photon energy. The possibility exists that these changes are a source of information on the ground state wavefunction, so it is vital that this area of interpretation be laid on a sound foundation. Progress in this area should result from experimental study over a wider range of photon energy, which is possible using synchroton radiation. Also theoretical studies of P.E. band intensities for this size of molecule should provide a challenging and rewarding problem.

Acknowledgements. I should like to thank my research students, especially Sally Jackson and Elaine Seddon, who have collaborated with me in these studies over the past ten years.

References

1. Furlani, C., Cauletti, C.: Structure and Bonding *35*, 119 (1978)
2. Cowley, A.H.: Prog. Inorg. Chem. *26*, 46 (1979)
3. Hamnett, A., Orchard, A.F.: Electronic Structure and Magnetism of Inorganic Compounds. London: Chem. Soc. 1972, Vol. 1, p. 1
4. Evans, S., Orchard, A.F.: Electronic Structure and Magnetism of Inorganic Compounds. London: Chem. Soc. 1973, Vol. 2, p. 1
5. Hamnett, A., Orchard, A.F.: Electronic Structure and Magnetism of Inorganic Compounds. London: Chem. Soc. 1974, Vol. 3, p. 218
6. Egdell, R.G., Potts, A.W.: Electronic Structure and Magnetism of Inorganic Compounds. London: Chem. Soc. 1980, Vol. 6, p. 1
7. Turner, D.W., Baker, C., Baker, A.D., Brundle, C.R.: Molecular Photoelectron Spectroscopy. London: Wiley Interscience 1970
8. Eland, J.H.D.: Photoelectron Spectroscopy. London: Butterworths 1974
9. Rabalais, J.W.: Principles of Ultraviolet Photoelectron Spectroscopy. New York: Wiley Interscience 1977
10. Brundle, C.R., Baker, A.D. (eds.): Electron Spectroscopy. London: Academic Press 1977, Vol. 1
11. Brundle, C.R., Baker, A.D. (eds.): Electron Spectroscopy. London: Academic Press 1978, Vol. 2
12. Brundle, C.R., Baker, A.D. (eds.): Electron Spectroscopy. London: Academic Press 1979, Vol. 3
13. Koopmans, T.: Physica *1*, 104 (1934)
14. Rohmer, M.-M., Veillard, A.: Chem. Comm. 250 (1973)
15. Coutière, M.M., Demuynck, J., Veillard, A.: Theoret. Chim. Acta. *27*, 281 (1972)
16. Connor, J.A., et al.: Mol. Phys. *28*, 1193 (1974)
17. Evans, S., et al.: J. Chem. Soc. Dalton 304 (1974)
18. Block, T.F., Fenske, R.F.: J. Amer. Chem. Soc. *99*, 4321 (1977)
19. Zenistra, J.D., Nienwpoort, W.C.: Inorg. Chim. Acta. *30*, 103 (1978)
20. Guest, M.F., et al.: Mol. Phys. *29*, 113 (1975)

21. Egdell, R.G., et al.: J. Electr. Spectr. *12*, 415 (1977)
22. Coleman, A.W., et al.: J. Chem. Soc. Dalton, 1057 (1979)
23. Cauletti, C., et al.: J. Electr. Spectr. *18*, 61 (1980)
24. Clark, J.P., Green, J.C.: J. Chem. Soc. Dalton 505 (1977)
25. Berndtsson, A., et al.: Phys. Scripta *12*, 235 (1975)
26. Samson, J.A.R., Gardner, J.L., Haddad, G.N.: J. Electr. Spectr. *12*, 281 (1977)
27. Cox, P.A.: Structure and Bonding *24*, 59 (1975)
28. Evans, S., Green, J.C., Jackson, S.E.: J. Chem. Soc. Farad. II *68*, 249 (1972)
29. Warren, K.D.: Structure and Bonding *27*, 45 (1976)
30. Cloke, F.G.N., Green, M.L.H., Morris, G.E.: Chem. Comm. 72 (1978)
31. Feltham, R.D., Sogo, P., Calvin, M.: J. Chem. Phys. *26*, 1354 (1957)
32. Prins, R., Reinders, F.J.: Chem. Phys. Lett. *3*, 45 (1969)
33. Henrici-Olivé, G., Olivé, S.: Z. Phys. Chem. (Frankfurt) *56*, 223 (1967)
34. Eischner, B., Herzog, S.: Z. Natur. *12a*, 860 (1957)
35. Dahl, J.P., Ballhausen, C.J.: Kgl. Danske Videnskub. Selskub. Mat. Phys. Medd. *33*, 5 (1961)
36. Bagus, P.S., Walgren, U.I., Almlof, J.: J. Chem. Phys. *64*, 2324 (1976)
37. Baerends, E.J., Ros, P.: Chem. Phys. Lett. *29*, 391 (1973)
38. Weber, J., et al.: J. Amer. Chem. Soc. *100*, 3995 (1978)
39. Shustorovich, E.M., Dyatkina, M.E.: Dokl. Chem. *131*, 113 (1960)
40. Schachtschneider, J.H., Prins, R., Ros, P.: Inorg. Chim. Acta *1*, 462 (1967)
41. Botrel, A., Dibout, P., Lissillour, R.: Theoret. Chim. Acta *37*, 37 (1975)
42. Green, J.C.: unpublished
43. Rösch, N., Johnson, K.A.: Chem. Phys. Lett. *24*, 179 (1974)
44. Kirschner, R.F., Loew, G.H., Müller-Westerhoff, V.T.: Theoret. Chim. Acta *41*, 1 (1976)
45. Turner, D.W.: Physical Methods in Advanced Inorganic Chemistry (Hill, H.A.O., Day, P., eds.) New York: Interscience, 1968
46. Evans, S., et al.: J. Chem. Soc. Farad. II *68*, 1847 (1972)
47. Rabalais, J.W., et al.: J. Chem. Phys. *57*, 1185 (1972)
48. Cauletti, C., et al.: J. Electr. Spectr. *19*, 327 (1980)
49. Evans, S., et al.: J. Chem. Soc. Farad. II *70*, 356 (1974)
50. Prins, R., Reinders, F.J.: J. Amer. Chem. Soc. *91*, 4929 (1969)
51. Prins, R.: Mol. Phys. *19*, 603 (1970)
52. Groenenboom, C.J., et al.: J. Organomet. Chem. *97*, 73 (1975)
53. Cloke, F.G.N., Green, M.L.H., Price, D.H.: Chem. Comm. 431 (1978)
54. Rettig, M.F., et al.: J. Am. Chem. Soc. *92*, 5100 (1970)
55. Almehnigen, A., Haaland, A., Samdal, S.: J. Organomet. Chem. *149*, 219 (1978)
56. Warren, K.D.: Inorg. Chem. *13*, 1243 (1974)
57. Cox, P.A., Evans, S., Orchard, A.F.: Chem. Phys. Lett. *13*, 386 (1972)
58. Cox, P.A., Orchard, A.F.: Chem. Phys. Lett. *7*, 273 (1970)
59. Anthony, M.T., Green, M.L.H., Young, D.: J. Chem. Soc. Dalton 1419 (1975)
60. Astruc, D., Green, J.C., Kelly, M.R.: unpublished
61. Fischer, R.D.: Theoret. Chim. Acta *1*, 418 (1963)
62. Clack, D.W, Warren, K.D.: Inorg. Chim. Acta *30*, 251 (1978)
63. Wild, D.J.: Part II Thesis; Oxford, 1975
64. Jackson, S.E.: D. Phil. Thesis; Oxford, 1973
65. Hayes, R.G., Edelstein, N.: J. Am. Chem. Soc. *94*, 8688 (1972)
66. Streitwieser, A., Müller-Westerhoff, U.: J. Am. Chem. Soc. *90*, 7364 (1968)
67. Downs, A.J., et al.: J. Chem. Soc. Dalton 1755 (1978)
68. Long, J. A.: Part II Thesis; Oxford, 1976
69. Clark, J.P., Green, J.C.: J. Organomet. Chem. *112*, C14 (1976)
70. Fragala, I., et al.: J. Organomet. Chem. *122*, 357 (1976)
71. Rosch, N., Streitwieser, A.: J. Organomet. Chem. *145*, 195 (1978)
72. Pygall, C.: Part II Thesis; Oxford, 1972

110

73. Green, J.C., Kanellakopoulos, B., Kelly, M.R.: unpublished
74. Reid, A.F., Scaife, D.E., Wailes, P.C.: Spectrochim. Acta 20, 1257 (1964)
75. Beatham, N., et al.: Chem. Phys. Lett. 63, 69 (1979)
76. Ballhausen, C.J., Dahl, J.P.: Acta Chem. Scand. 15, 1333 (1961)
77. Green, J.C., Green, M.L.H., Prout, C.K.: Chem. Comm. 421 (1972)
78. Green, J.C., Jackson, S.E., Higginson, B.: J. Chem. Soc. Dalton 403 (1975)
79. Petersen, J.L., Dahl, L.F.: J. Am. Chem. Soc. 96, 2248 (1974)
80. Petersen, J.L., et al.: J. Am. Chem. Soc. 97, 6433 (1975)
81. Lauher, J.W., Hoffmann, R.: J. Am. Chem. Soc. 98, 1729 (1975)
82. Fragala, I., Ciliberto, E., Thomas, J.L.: J. Organomet. Chem. 175, C25 (1979)
83. Green, J.C., Payne, M., Teuben, J.: unpublished
84. Green, J.C.: unpublished
85. Condorelli, G., et al.: J. Organomet. Chem. 87, 311 (1975)
86. Clark, J.P., Green, J.C.: J. Less Common Met. 54, 63 (1977)
87. Green, J.C., Kelly, M.R., Lappert, M., Yarrow, P.: unpublished
88. Fragala, I., et al.: J. Organomet. Chem. 120, C9 (1976)
89. Evans, S., et al.: J. Chem. Soc. Farad. II 68, 905 (1972)
90. Evans, S., Green, J.C., Jackson, S.E.: J. Chem. Soc. Farad. II 69, 191 (1973)
91. Lappert, M.F., Pedley, J.B., Sharp, G.: J. Organomet. Chem. 66, 271 (1974)
92. Gayler, L., Wilkinson, G., Lloyd, D.R.: Chem. Comm. 497 (1975)
93. Green, J.C., et al.: J. Chem. Soc. Dalton 1403 (1978)
94. Cradock, S., Savage, W.: Inorg. Nucl. Chem. Lett. 8, 753 (1972)
95. Evans, S., et al.: Diss. Farad. Soc. 54, 112 (1969)
96. Lichtenberger, D.L., Fenske, R.F.: Inorg. Chem. 13, 486 (1974)
97. Higginson, B.R., et al.: J. Chem. Soc. Farad. II 71, 1913 (1975)
98. Hall, M.B.: J. Am. Chem. Soc. 97, 2057 (1975)
99. Hall, M.B., Fenske, R.F.: Inorg. Chem. 11, 768 (1972)
100. Guest, M.F., Hall, M.B., Hillier, I.H.: Chem. Phys. Lett. 15, 592 (1972)
101. Guest, M.F., Hall, M.B., Hillier, I.H.: Mol. Phys. 25, 629 (1973)
102. Guest, M.F., et al.: J. Chem. Soc. Farad. II 71, 902 (1975)
103. Elian, M., Hoffmann, R.: Inorg. Chem. 14, 1058 (1975)
104. Cradock, S., Ebsworth, E.A.V., Robertson, A.: J. Chem. Soc. Dalton 22 (1973)
105. Behan, J., Johnstone, R.A.W., Puddephatt, R.J.: Chem. Comm. 444 (1978)
106. Green, J.C., Jackson, S.E.: J. Chem. Soc. Dalton 1698 (1976)
107. Symon, D.A., Waddington, T.C.: J. Chem. Soc. Dalton 2140 (1975)
108. Lichtenberger, D.L., Fenske, R.F.: J. Am. Chem. Soc. 98, 50 (1976)
109. Basso-Bert, M., et al.: J. Organomet. Chem. 136, 201 (1977)
110. Lloyd, D.R., Schlag, E.W.: Inorg. Chem. 8, 2544 (1969)
111. Baerends, E.J., Oudshoorn, C., Oskam, A.: J. Electr. Spectr. 6, 259 (1975)
112. van Dam, H., Oskam, A.: J. Electr. Spectr. 16, 307 (1979)
113. van Dam, H., Oskam, A.: J. Electr. Spectr. 17, 357 (1979)
114. Hill, W.E., et al.: Inorg. Chem. 18, 2029 (1979)
115. van Dam, H., Tepstra, A., Stufkens, D.J., Oskam, A.: in press
116. Flamini, A., et al.: J. Chem. Soc. Dalton 695 (1978)
117. Hall, M.B., et al.: Mol. Phys. 30, 829 (1975)
118. Connor, J.A., et al.: Mol. Phys. 31, 23 (1976)
119. Green, J.C., Powell, P., van Tilborg, J.: J. Chem. Soc. Dalton 1974 (1976)
120. Green, J.C., Kelly, M.R.: unpublished
121. Green, J.C., van Tilborg, J.: unpublished
122. Dewar, M.J.S., Worley, S.D.: J. Chem. Phys. 50, 654 (1969)
123. Worley, S.D.: Chem. Comm. 980 (1970)
124. Dewar, M.J.S., Worley, S.D.: J. Chem. Phys. 51, 1672 (1969)
125. Jackson, S.E.: D. Phil. Thesis; Oxford, 1963
126. Rohmer, M.-M., Demuynck, J., Veillard, A.: Theoret. Chim. Acta 36, 93 (1974)
127. Veillard, A.: Chem. Comm. 1022, 1427 (1969)

128. Brown, D.A., Owens, A.: Inorg. Chim. Acta *5*, 675 (1971)
129. Lloyd, D.R., Lynaugh, N.: Electron Spectroscopy (Shirley, D.E., ed.) North-Holland, Amsterdam, 1972, p. 445
130. Batich, C.D.: J. Am. Chem. Soc. *98*, 7585 (1976)
131. Seddon, E.A.: D. Phil. Thesis; Oxford, 1980
132. Gower, M., et al.: J. Chem. Soc. Dalton 316 (1977)
133. Lichtenberger, D.L., Fenske, R.F.: J. Am. Chem. Soc. *98*, 50 (1976)
134. Lichtenberger, D.L., Sellman, D., Fenske, R.F.: J. Organomet. Chem. *117*, 253 (1976)
135. Whitesides, T.H., Lichtenberger, D.L., Budink, R.A.: Inorg. Chem. *14*, 68 (1975)
136. Ulman, J.A., Anderson, E.L., Fehlner, T.P.: J. Am. Chem. Soc. *100*, 456 (1978)
137. Evans, S., et al.: J. Chem. Soc. Farad. II *70*, 417 (1974)
138. Rankin, R.: unpublished

Vibrations in Interaction with Impurities

Robert Englman

Soreq Nuclear Research Centre, Israel Atomic Energy Commission, Yavneh, Israel

A review is presented of the effects of interaction between electronic states on an impurity ion or molecule and the vibrations of the host. Primarily, interactions linear in the vibrational amplitude are treated for their static and time-dependent effects, while mass or force-constant defects are considered only to the extent of bringing out the analogies and distinctions between them and linear interactions. Some emphasis is placed on degenerate impurity states and on the phase-correlation in the electronic and vibrational motions. We describe in some detail experimental and theoretical works which throw light on the character of the vibrational motion that is in interaction with impurity electrons. We conclude that this can vary with circumstances, between extremely localised, molecular type and extended, phonon-like behaviours, where large energy exchange (as e.g. optical) phenomena belong to the former and small energy exchange or scattering experiments to the latter.

Table of Contents

1.1 Introduction

Following ideas developed in the 1910's [25,38] the motions of atoms or ions in solids have been analysed in terms of small oscillations about their equilibrium positions. These oscillations or vibrations have thus become a concept of central importance in the theory of solids. They are responsible for the thermal properties in insulating solids and, partly, in conductors as well as for several other properties of which we shall here name only the mechanical and acoustic ones. Description of these topics are available in several modern textbooks on solid state physics [47,80,92].

The subject of this Report is also vibrations in solids, but the emphasis here is essentially different from that in the previously quoted sources and, indeed, in most of the works dealing with "lattice-vibrations", "phonons", "elastic-waves", etc. First, the solids that are of concern to us are impure. That is, in addition to the host constituents, they contain atoms, ions or molecules. For many purposes the concentration of these is supposed to be so low that interactions between them can be neglected and so the mental picture that we shall operate with is that of a single impurity in an infinite lattice. In practice, impurity/host concentration ratios of less than 10^{-4} conform to this picture. In the upper range of this limit, say in the $10^{-4}-10^{-5}$ region, one may find that pairs of impurities can form (e.g. Si_2 in Ge [54]; N_2 in GaP [5,53]; Bi and a donor in GaP [37]; F_2-centres in alkali halides [71,77]; Br_2^{2-} in KI [159]) but since those pairs are still impurities, our considerations apply to them too. One phenomenon needs however to be excluded from the single impurity picture that we propose. This is energy-exchange which can take place even between impurities of low, sub-10^{-5} concentration. We shall not treat energy-exchange (summaries [118,119,129,163,168]) even though the treatments of vibration that we expound here have also applications in that field.

The single impurities have electronic properties whose details are important for us and which will be treated in this review on an individual basis. In contrast, the electronic structure of the host lattice is of no interest and will not appear explicitly in this discussion, even though (indirectly) it is responsible for the vibrational properties of the lattice. Let us now state the basic tenet underlying our treatment: The vibrations of the atoms in the lattice interact with the electrons on the impurity and affect its properties (energy levels, wave functions, transition probabilities, etc.). Before formulating the situation in general terms let us fix our ideas by quoting some characteristic examples.

1.2 Vibration Effects in Optical Spectra

The interplay between vibrations and electronic states will be first illustrated by a somewhat novel and unusual system, in which the vibrations do not show their presence in an obvious, explicit way but through their effect on the dynamic properties of the impurity electrons. Figure 1 shows absorption due to the ionic impurity Cu^+ in the intrinsically transparent LiCl crystal [139]. Four bands are apparent of which I and III are attributed to Cu^+ pairs. Band II is interpreted as due to the transition $d \rightarrow s$, Laporte-rule forbidden on $Cu^+(Cl^-)_6$ which is made E1-allowed by the inter-

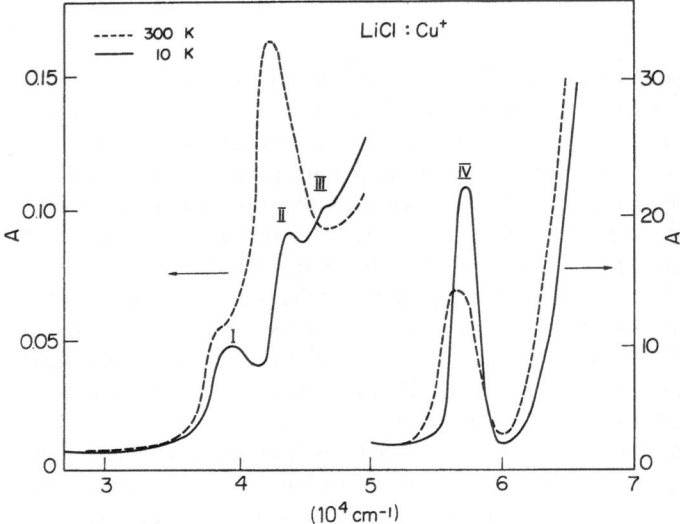

Fig. 1. Absorption strength A for LiCl:Cu$^+$ at two temperatures as a function of wave-number. (A is given as 2.9×10^{-4} times the molar extinction coefficient) (Simonetti and McClure 1977[139)]) The four bands are discussed in the text

mediation of a vibration of odd symmetry. Though not apparent in the structure of the band its effect is felt in the variation of the band intensity I(T) with temperature T according to the relation

$$\frac{I(T)}{I(0)} = \coth\,(\hbar\omega/2\,KT)$$

A vibrational energy of $\hbar\omega = 100\,\text{cm}^{-1}$ fits the data well (Table 1).

The strong band IV is caused by an E1-allowed d → p transition. No inducing vibration is needed here, but the change in the electronic configuration causes a displacement in the vibrations. The width of the band is a measure of the relaxation following excitation and from its low temperature value of about $1500\,\text{cm}^{-1}$ one can

Table 1. Properties of some absorption bands in LiCl:Cu$^+$ (after [139)]

Band	Assignment $^1A_{1g}(3d^{10}) \to$	Temperature in °K	Mean energy in cm^{-1}	Width in cm^{-1}	Evident vibrational effects
II	$^1E_g(3d^94s)$	10	43620	2200	Temperature caused intensity-increase and peak-shift
		300	43010	2500	
IV	$^1T_{1u}(3d^94p)$	10	57310	1500	Temperature caused peak-shift and width-enhancement
		300	56750	3700	

estimate the Huang-Rhys number S (which is a measure of the electron-vibration coupling strength) at 15, if no allowance is made for the inhomogeneous width. Other vibrational effects are the temperature enhanced width and the shift to the red by about 800 cm^{-1} at room temperatures (in the presence of anharmonicity[50a)]). These features are regular vibrational or phonon effects in d^n-ion impurities and also in other point defects.

The nature of the vibrations that are in play here is not directly apparent and theoretical investigations, of which an account will be given in the sequel, are aimed to elucidate with the aid of related experimental knowledge the character of the vibrations.

1.3 Resonances in Thermal Conductivity

For our second example we choose the ionic impurity Ni^{3+} inserted in the trigonal crystal Al$_2$O$_3$. This is, by way of contrast to the previous case, a thoroughly studied system whose properties were observed by ESR[62,137a)] including also electric field[58)] – and stress-induced effects[137b)], ultrasonic attenuation[64)], Raman spectroscopy[31)] and temperature-dependent thermal conductivity measurements[131,134)]. It is the last type of observation that focuses attention in the most direct way to the subject of this report: the vibrational motion engendered by the impurity. The experimental curves result, shown in Fig. 2, indicate the presence in the phonon-scattering mecha-

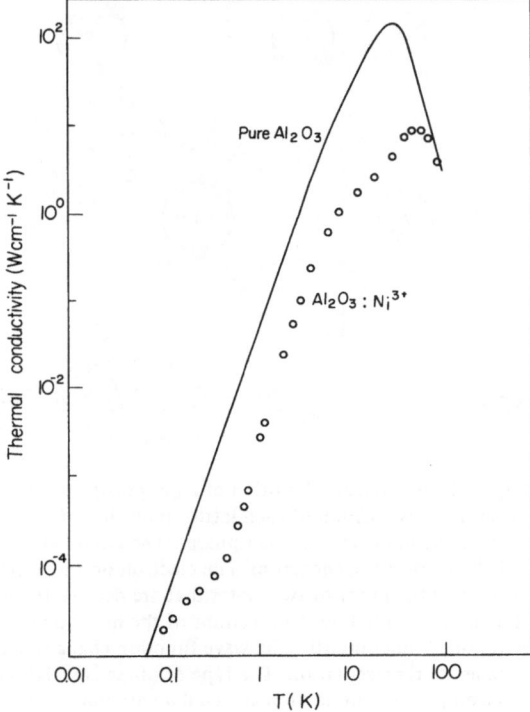

Fig. 2. Thermal resistance of Ni^{3+} ions in Al$_2$O$_3$ (Salce and de Goer 1979[134)]). The full curve follows the experimental points (not shown) for pure corundum. The circles are measurements on nickel-containing-Al$_2$O$_3$ (less than 20 ppm by weight). The deviation of the two data sets arises from the scattering of phonons by the Ni impurities. Two dips near 0.7 K and 20 K are identifiable

Thermal conductivity (Wcm^{-1} K^{-1})

Pure Al$_2$O$_3$

Al$_2$O$_3$: Ni$_i^{3+}$

T(K)

nism[91] of two impurity-modes (at 0.74 and 55 cm^{-1}) by the dips in the impure crystals conductivity curve in the region of 0.7 and 20 K. The modes in this second example are not pure vibrational excitations but rather concerted motion of the electronic and vibrational degrees of freedom.

From the fitting of the various experimental data to theory[1,15] it emerges that the $Al_2O_3:Ni^{3+}$-system has an orbitally degenerate ^2E ground state and is in its trigonal environment subject to a Jahn-Teller type instability[50b,66,147]. The instability arises from the co-degenerate states of the centre that have lower than trigonal symmetry (Fig. 3) and which exert a force on the lattice lowering its symmetry from

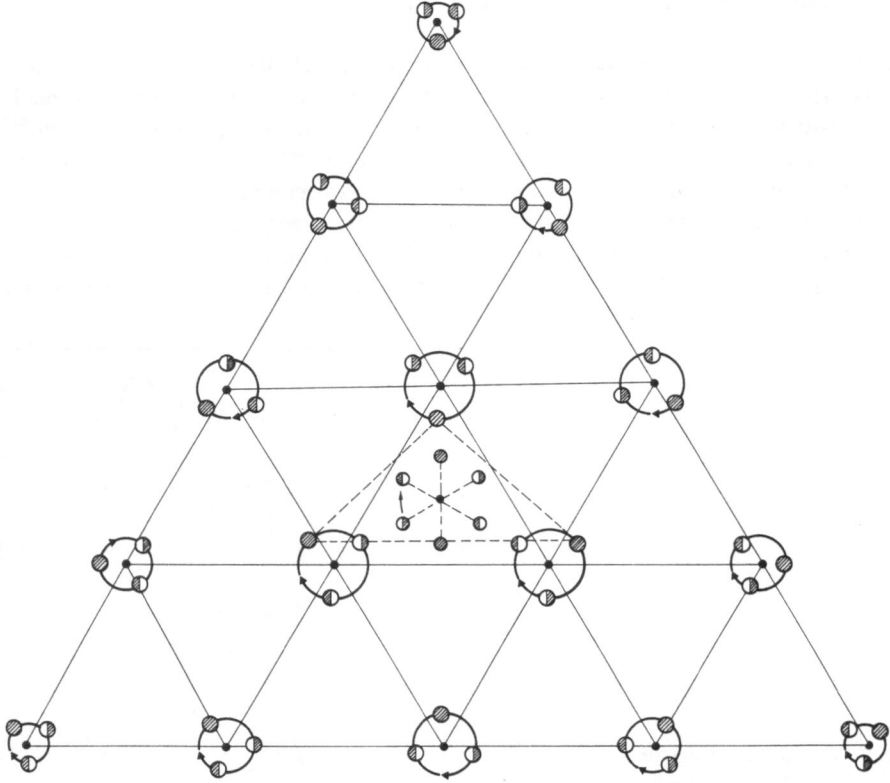

Fig. 3. Phase-concerted motion of an impurity-electron and vibrations in a trigonal lattice. The instantaneous position of each lattice point lies on the circle surrounding the point, with the times being indicated by lunar phases. The circles are drawn progressively smaller as one goes radially outwards (schematic). The electron on the impurity (centre of the figure) is taken as occupying the lower of two states that are degenerate in trigonal symmetry. The motion of the state is represented by the positions of the nodal plane (shown by broken lines) at various stages. After one "lunar month" the wave function changes sign, this being a peculiarity of the doubly degenerate representation. The type of phase correlation shown in this figure is absent in the electron-phonon interaction associated with mass or force-constant changes at a defect

the trigonal. The details of this situation need not be spelled out in this Introduction, but the following points are worthy of notice since they throw light on the subject of this report in general. An interaction is involved between electrons localised on Ni^{3+} and the lattice displacements. This is already the second instance of such an interaction, and differs from the previous one in that the geometry of the lattice rather than the dynamics of an electronic process is primarily affected by the interaction. The formal treatment of Sect. 3.2 will introduce the interaction in a general form that will include these cases and several other ones.

One can also note the "dual" character[117,152] of the instability. This has a *localised* aspect, since it originates on the localised states of the impurity centre, as well as an *extended* aspect since the electrons interact with phonons, the quantised motion spreading out all over the lattice. Clearly the quantitative effects of the interaction will depend on the properties of the phonons: for example, one can safely anticipate that the effects will be small if the lattice resists well to pressure (of the localised electrons), or, equivalently, if the phonon energies are high.

There are a number of areas where the electron-phonon interaction is of central importance to the phenomenon: E.g. spin-phonon interactions in paramagnetic and nuclear magnetic relaxations[155], nucleation of phase transitions in impure crystals[78], optical activity[72,84], tunelling (as e.g. between off-centre positions of impurity ions[96]).

This report will be relevant to all these fields but will not cover them. Let me explain the second half of this statement. In a broad way one can say that in each case of the phenomena listed above one has an electronic part, a vibrational part and an interaction between these. The Hamiltonian of the formalism can also be broken up into three corresponding terms. We shall devote one section (see 2.4) to the physical origin of the interaction term and the rest will be about vibrations, in the context of their interaction with impurity electrons, but there will be no attempt to delve into the electronic part. This scheme of study was implied in our title which has vibrations as the subject of this report. This approach of ours to the interplay of electronic and vibrational degrees of freedom in impure insulators has a biased tinge and we present now a justification for it.

1.4 The Problematics of Locally Interacting Vibrations

In an ideal, perfect crystal energy is transported by lattice waves of uniform amplitude, called pure phonons. In non-crystalline but macroscopically homogeneous materials (like glasses or rubber) phonons have individually non-uniform amplitudes (shaping themselves to the local non-uniformity of the crystal) but certain averages of phonons relevant to energy transport are still spatially uniform. Thus the situation in these materials is phenomenologically not much different from that in a perfect crystal and what we shall have to say about crystals will also hold in some vague, less-precise way for glasses.

If a local impurity is introduced in a crystal most of the phonons will undergo a minor and essentially insignificant adjustment as regards their energies and spatial distributions. More fundamental changes can also occur in a small number of vibra-

119

tional modes and there will appear two new types of modes called localised and reso-nance modes[36,82,140,148].

Localised modes are stationary vibrational modes of the crystal, that peak near the impurity and die-off away from it. They arise in certain cases due to changes in the local elastic coefficient or in the mass density of the crystal in the vicinity of the impurity. These changes also give rise to resonance-modes, whose amplitudes too de-crease on going away from the impurity centre, but there is this difference between these and the localised modes that the former are not stationary modes. Instead they represent excitations that start near the impurity and then diffuse away into the crys-tal without returning to the impurity.

Although these two types of modes both arise due to impurity-vibration inter-actions, for reasons that will be made clear later our main interest lies not in these but in the phase-matched motion of the localised electron and of the lattice phonons. This type of motion will occur independently of whether there are or are not localised or resonance modes.

Figure 3 is shown to help visualise the phase-concerted electron-phonon inter-action, in a case that involves two electronic states and associated excitations of the vibrations. Although the figure is no substitute for a rigorous argument (this will be given in Sect. 4.4.3), intuitive reasoning aided by the figure suggests that the phase-concerted electron-phonon coupling generates vibrational excitations in the vicinity of the impurity. Then there arises the following dilemma: shall we represent these excitations in terms of *phonons*[61,160] or in terms of *local modes* similar in nature to resonance-modes? The first alternative ensures that the modes excited by the elec-tronic motion be stationary but does not do justice to the local character of the ex-citation. The second choice works in the opposite sense. It represents the modes inter-acting with the local electrons as themselves local; however, these modes, like the resonance-modes mentioned earlier, cannot hold the excitation to the impurity local-ity for indefinitely long.

Even more important than the somewhat philosophical question of how to view these excitations is the practical consideration: Which approach yields reliable quan-titative results more readily? Since some of the problems are mathematically quite complex, the practical aspect carries considerable weight. One might also add that even if one prefers the local mode description of the interacting vibrations, there are various manners by which this local mode can be constructed. Some of these ways have been reviewed by Halperin and the author[51].

Having posed the questions that appear to lie at the root of our subject, we should say that different problems and situations have led to different answers and approaches. One can also safely add that it was the different backgrounds and attitudes on the parts of the researchers that have resulted in the variety of methods. By now these amount to quite a number. In consequence, we feel that treatments of the vibrations interacting with electrons on an impurity are in need of a summary.

1.5 Experimental Background

The subject of electron-phonon interaction has strong experimental connection, to which it owes its development. Line-widths in ESR (Electron Spin Resonance) provide information on the strength of that interaction and so do, after subtraction of inhomogeneous contribution, the widths of line-like optical spectra and of the broad optical and ultra-violet bands due to, e.g., colour centres or ionic impurities. The intensity of the Mössbauer-line (the Debye-Waller factor) and its optical analogue, the zero-phonon line-strength decrease in relation to the strength of the interaction. The vibronic line-structure, i.e. the vibrational excitations accompanying electronic transitions, are due to the same interaction whether this be to electronically degenerate states (the Jahn-Teller effect) or to non-degenerate ones. The impurity caused scattering of sound and ultra-sonic waves and the thermal conductivity of impure insulators also arise from the same mechanism.

It is thus evident that potentially a wealth of data relates to the phonon-electron coupling. It seems appropriate to classify this under three headings, according to whether

(a) one investigates an essentially electronic quantity and measures the effect of the coupling on this (e.g. the temperature dependence of the zero phonon line-width),

(b) one tests the system with a vibration-like probe (as, e.g., in acoustic-wave scattering) and thus investigates the electron-phonon contribution to the result or

(c) one uses a technique that locks on a property that is itself the outcome of the electron-phonon coupling.

Example of this last class is the infrared spectroscopy of electronically degenerate systems. Here the energy-level structure arises from the vibronic (vibrational and electronic) coupling and the separation of levels into electronic or vibrational ones is meaningless. (E.g. MgO:V^{2+} [7,146]).

Though the amount of experimental data related to or arising from the electron-phonon interaction is very high, still, only a small part of this data is relevant to the subject of this study, which is about the nature of vibrations coupled to localised electrons. Thus we are seeking experimental indications (or justification for theoretical predictions) how these vibrations are distributed in space, how they develop in time and what is their break-down in energy or momentum-space. A mapping of the vibrational amplitudes near the impurity can be achieved variously, e.g. by ENDOR (Electron-Nuclear Double Resonance); regrettably, no experiments have been performed yet. Time-resolved spectroscopy in the pico-second range or shorter is an obvious tool to probe the life-time of the excited vibronic levels. Hot luminescence (i.e. emission from these levels [128]) will occur only before the diffusion of the excitation from the impurity centre. A further possibility that has come into use comparatively recently is the observation of phonon pulses arising from excited vibrational levels [42,102,130]. The investigation of the energy distribution of the coupled electronic-vibrational motion is relatively the simplest. Positions of the peaks in the optical spectra reveal the energy regions in question. Critical points in the vibrational spectra (discussed in the next section, (Fig. 4)) and localised vibrations due to impurities have been identified in semiconductors [26,70,81,110–112,141] and in insulators [104,138,161,167].

2.1 The Oscillations of the Lattice

A brief introduction, amounting to hardly more than definitions to the concepts of vibrations in the lattice will be provided here. More through information is available in several texts [52,99].

2.2 Vibrations of the Host

First we suppose that the crystal is regular (without defects or impurities) and that the electronic states in it are fixed. The latter assumption enables us to treat the vibrations within the Born-Oppenheimer approximation. The potential energy of the lattice is obtained from the potential energy function of all particles (electrons and atoms) as the average in the fixed electron states. It can be expanded in the displacement vectors $u_\alpha(l\kappa)$ ($\alpha = x, y, z$) of the κ-type atom situated at the l'th lattice cell with respect to its equilibrium position $X_\alpha(l\kappa)$. We write the potential energy of the crystal at an arbitrary point of the lattice as $V(\{l\kappa\})$ and as $V(\{X\})$ at the equilibrium positions. Then

$$V(\{l\kappa\}) - V(\{X\}) = \tfrac{1}{2} \sum_{\substack{\alpha\beta \\ ll'\kappa\kappa'}} \left[\frac{\partial^2 V(\{l\kappa\})}{\partial u_\alpha(l\kappa)\, \partial u_\beta(l'\kappa')} \right]_X u_\alpha(l\kappa)\, u_\beta(l'\kappa') \\ + \text{terms of higher order in } u \qquad (2.1)$$

The kinetic energy is

$$K = \tfrac{1}{2} \sum_{\alpha l\kappa} M_\kappa\, \dot{u}_\alpha^2(l\kappa)$$

where the dot denotes time differentiation and M_κ is the mass of the κ-type atom. The higher order terms in (2.1) are essential in strongly anharmonic crystals [12,33,74]. Restricting our treatment to terms harmonic in the potential, we can transform the displacements to normal coordinates as follows

$$Q_j(k) = N^{-\frac{1}{2}} \sum_{\alpha l\kappa} M_\kappa^{\frac{1}{2}}\, e_\alpha^*(\kappa|k_j)\, e^{-ik\cdot X(l\kappa)}\, u_\alpha(l\kappa) \qquad (2.2)$$

where N is the number of lattice points, k the wave-rector, j is the branch index and e the polarisation. The transformation brings the potential to a canonical form diagonal in $Q_j(k)$

$$H_{phonon} = \tfrac{1}{2} \sum_{kj} \omega^2(kj)\, Q_j^*(k)\, Q_j(k) \qquad (2.3)$$

in which the eigenfrequencies $\omega(kj)$ appear.

The kinetic energy of the normal modes stays diagonal:

$$K = \frac{1}{2} \sum_{kj} |\dot{Q}_j(k)|^2 \qquad (2.4)$$

In a non-crystalline solid (whose lattice is structurally or compositionally highly irregular) one also has the normal modes and the diagonal form (2.3), but the mode labels kj would be replaced by some other ones, having no clearcut physical or geometrical meaning. In a nearly harmonic crystal the anharmonic part in (2.1) causes the normal modes to have finite life-times[89,98]. In stable lattice, i.e. well below melting or other phase-transition points these are characteristically of the order of some hundreds inverse frequencies.

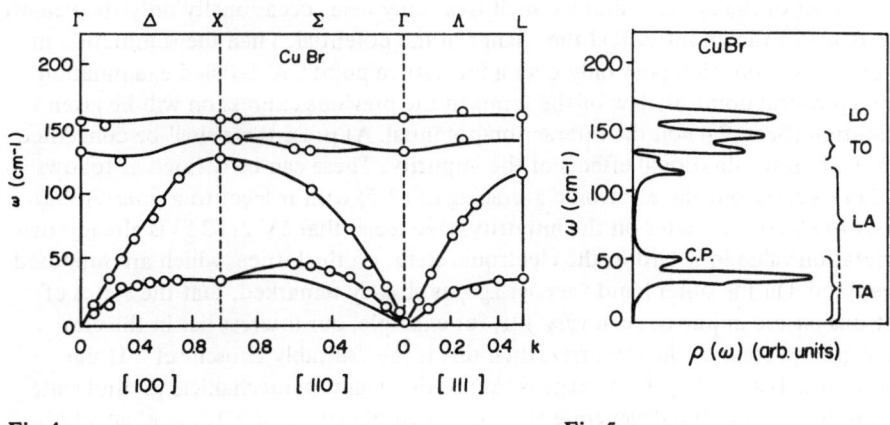

Fig. 4   Fig. 5

Fig. 4. Vibrational frequencies of a diatomic lattice. CuBr is taken as an example. Shown are experimental results (circles) due to neutron diffraction by Prevot et al.[125] and theoretical curves from calculations of Vardeny et al.[157]. Eigen-frequencies along three principal directions in reciprocal-space k are shown. Points of high symmetry appear above the figure

Fig. 5. Density of vibrational levels ρ as function of frequency ω. The vibrational types (longitudinal L and transverse T, acoustic A and optic O) are indicated on the right. The region of a critical point (C.P., where the ω vs. k curve in Fig. 4 is flat) is also shown. (After Vardeny et al.[157])

The normal mode frequencies $\omega(kj)$ form for each branch-index j a quasi-continuous band as function of the energy, with gaps (forbidden regions) between different bands (Fig. 4). The deduced curve for the density of vibrational levels $\rho(\omega)$ (= number of vibrational modes in a crystal of unit volume per unit energy) appears in Fig. 5. Sharp peaks of ρ on the density of states represent the critical points on the ω vs. k curve, i.e. the points where this curve is flat or nearly so[90] (Chap. 4, Sect. 4).

2.3 Vibrations of the Impure Crystal

The insertion of an impurity will change the potential energy of the crystal. The change δV in the potential energy that comes from the impurity electrons (coordinates r_i) can be expanded as before in the displacements $u_\alpha(l\kappa)$ of the lattice atoms from their standard positions X:

$$\delta V \equiv \delta V(r_i, \{l\kappa\}) = \delta V(r_i, X(l\kappa)) + \sum_{\alpha l\kappa} \left[\frac{\partial \delta V}{\partial u_\alpha(l\kappa)} \right]_X u_\alpha(l\kappa)$$

$$+ \frac{1}{2} \sum_{\substack{\alpha\beta \\ ll\kappa\kappa'}} \left[\frac{\partial^2 \delta V}{\partial u_\alpha(l\kappa)\, \partial u_\beta(l'\kappa')} \right]_X u_\alpha(l\kappa)\, u_\beta(l'\kappa') + \ldots$$

(2.5)

In most of the systems that we shall treat only near, occasionally only the nearest, neighbours of the impurity feel the change in the potential. Then the summation in the previous expression goes only over a few lattice points. A detailed examination from a physical point of view of the terms in the previous expression will be given in the section (Sect. 2.4) on the interaction potential. At present we shall be concerned with the purely vibrational effects of the impurity. These can be defined as follows:

Let us carry out the electronic averaging of (2.5) with respect to a *suitably chosen* set of electronic states on the impurity. (We recall that δV in (2.5) is already the expectation value in regard to the electronic states on the lattice, which are supposed to be fixed. On the other hand we envisage, as already remarked, that the states of electrons on the impurities can vary.) If, for example, our interest lies in the electronic ground state of the impurity, then this is the "suitably chosen set". If the ground state is a singlet, the average is simply the quantum mechanical ground state expectation value; for a degenerate state with components e (e = 1, ... , g) an additional averaging of the individual expectation values δV_{ee} required. For non-degenerate states each electronic state e will be weighted with a probability factor

$$p_e \ (e = 1, \ldots , g) \,.$$

(2.6)

For a static situation and in thermal equilibrium each electronic state has to be weighted by its Boltzmann factor, but in other, non-static situations other weights are preferable. Thus for optical transitions between states (of different energies) equal weighting of the initial and final level is recommended. (Sect. 3.1 contains a more detailed discussion of the proposed averaging of δV).

The standard positions $X_\alpha(l\kappa)$ of the impure crystal will now be so redefined that in the electronic average of (2.5), namely

$$\overline{\delta V} = \sum_e p_e\, \delta V_{ee}$$

(2.7)

there be no linear term. In this way we allow for a static distortion of the lattice surrounding the impurity. The average quadratic term in (2.5) cannot be made to vanish, in general and will be written as δV_2.

The presence of impurity masses will change the kinetic energy in (2.2) and this will be denoted by

$$\delta K = \frac{1}{2} \hbar^2 \sum_{\alpha l \kappa}' \delta\left(\frac{1}{M_\kappa}\right) \frac{\partial^2}{\partial u_\alpha^2 (l\kappa)} \tag{2.8}$$

the summation being restricted to the impurities.

The sum of δV_2 and δK defines a matrix in the space of the displacements. This matrix is nonzero, or significantly so, only for a few displacements closely surrounding an impurity. These form what is termed, following Koster and Slater[90,99], Maradudin et al. the defect or impurity space. Green's function or other equivalent methods[46,48,89,149,170] enable one to find the characteristic frequencies ω' of oscillations due to the impurity. In the simplest case of an isotopic mass defect (M') in a monatomic lattice of atomic mass M the equation yielding ω' can be written as

$$\frac{M' - M}{3 NM} \omega'^2 \sum_{kj} [\omega^2(kj) - \omega'^2]^{-1} = 1 \tag{2.9}$$

where the summation is over all normal modes of the host.

Solutions ω' of (2.9) which lie between the bands or above them correspond to localised (or impurity) modes that are stationary modes of vibrations[106]. When the roots of (2.9) fall inside a band, a resonant mode arises whose amplitude (like that of a localised mode) falls off away from the impurity, but which is not stationary[27,82,148]. Since modes having these properties will also arise from the linear electron-phonon interaction (Sec. 4.4) we explain their physical meaning here, with a view also to later relevance.

Imagine a situation in which the impurity potential is switched on from an initial value of zero. Excitation of the resonant modes (and also of the localised modes, if there are any) will occur, setting in vibrational motion the atoms near the impurity. This excitation will subsist for some time but will eventually spread out to the farther reaches of the lattice, the life-time of the localisation being dependent on the density of levels $\rho(\omega')$, but characteristically being several multiples of $(\omega')^{-1}$, or $10^{-11} - 10^{-12}$ seconds. Therefore for very fast physical phenomena (below picoseconds or so), the resonant modes will show the behaviour of bona-fide stationary modes, whereas for longer-lived occurrences it will appear as a transient. Light atom substituents generate localised modes. Those arising from hydrogen or its isotopes were frequently detected, e.g. in KBr[104], etc.[135], while F^-, OH^-, Sm^{2+} or Tl^+ in the same host were observed by Ward and Timusk[161]. Absorption lines due to Na^+ in AgBr[69], Ag^+[138] and the rare-earth ions Eu^{2+} and Yb^{2+}[101] in KI have been identified as belonging to localised modes. Resonant Raman scattering from F centres in NaCl and KCl showed localised as well as resonance modes[56,167].

Infra-red absorption peaks due to resonance modes of NaCl:Cu^+[164] were found to agree with the interpretation of the dips in the thermal conductivity vs. temperature curves[28]. Such effects on the thermal conductivity were earlier found by Walker and Pohl[158] for KCl:I^-. Resonance modes of the system NaCl:Ag^+ were detected

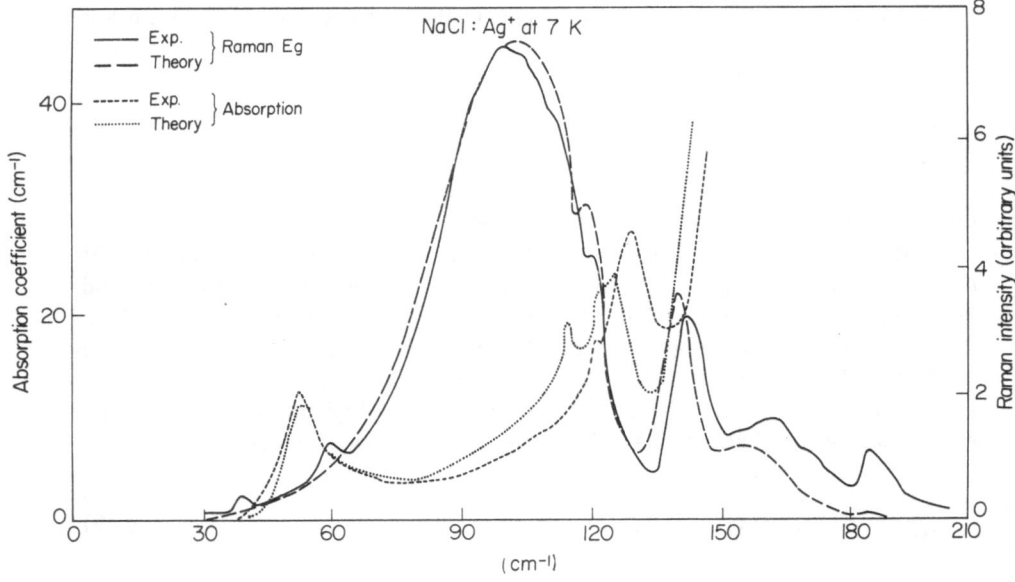

Fig. 6. Low temperature spectra of NaCl : Ag⁺. Raman (right hand scale) and absorption spectra (left hand scale) are shown. (After Montgomery et al. [105])

both by infra-red and Raman techniques [105] and with Eu^{2+} in SrF_2 and BaF_2 [32] or OH^- in KBr [122] (Fig. 6).

The theory of vibrational structure due to localised impurity modes in absorption or emission spectroscopy is essentially the well-known Pekar-Huang-Rhys theory developed in the 50's [75,123,124]. On the time scale of optical transitions resonance modes follow the same rules, with an additional broadening mechanism, due to their finite lifetimes, contributing to the widths. Cross sections for Raman scattering due to impurity modes were calculated with several models (mainly in the harmonic approximation) by, e.g., Ashkin [6], Trifonov and Peuker [154], Nguyen Xuan Xinh [113]. Anharmonic effects notably increase the complexity of the results; in addition they provide some new features [4] in infra-red absorption, Raman scattering or thermal conductivity (KBr:Li⁺ [16]). For the molecular impurity C_3 in Ne [23] vibration-phonon coupling increased the vibrational level width.

Several review articles deal extensively with the absorption and Raman spectroscopic manifestations of localised or resonance modes in impure crystals [89,140,149]. The experimental status is well documented in these and particularly in a recent compendium [9]. It is recalled that the emphasis of these reviews is different from that in this Report. They treat effects due to mass or force constant defects in the crystal. These are essentially averages over a vibrational period of the mode. In opposition, our concern is with effects coming from the phase-concerted interaction (linear in the vibrational coordinate). We turn to this in the next section, though for brevity's sake we shall not continue to carry the adjective "phase-concerted" in specifying the linear interaction.

2.4 Linear Electron-Vibration Coupling

2.4.1 General Results

A brief summary is here attempted of the forms of the linear electron-phonon interaction that have been used, with emphasis on the spatial variation of the interaction or equivalently on the wave-vector (k) dependence of its Fourier-transform.

The interaction between a free electron in a metal and the ionic displacements u is frequently given by the deformable potential approximation (E.g.[92], p. 128). If the potential felt by the electron in position r is $V(r)$ without the lattice being distorted, then upon distortion of the lattice to an extent of $u(r)$, the potential experienced by the electron is $\mathscr{V}(r-u)$. Another approximation views the potential as the sum of separate ionic potentials based on the instantaneous positions $X(l\kappa) + u(l\kappa)$ of each ion

$$\mathscr{V}(r) = \sum_{l\kappa} V[r - X(l\kappa) - u(l\kappa)]$$

and this form already contains the effect of ionic displacements. A comparison of this, the rigid-ion approximation, and the preceding one was made by Bardeen [8]. The spatial behaviour of the linear electron-vibration coupling (the u-derivative of the potential) will depend on the space dependence of the original potential.

The interaction in non-metals (e.g. ionic crystals or covalent semiconductors) will be now expressed in terms of phonon creation $a_{\kappa j}^*$ and destruction $a_{\kappa j}$ operators related to the pnonon amplitudes in (2.2) by

$$a_{kj} = \left[\frac{\omega(kj)}{2\hbar}\right]^{1/2} [Q_j(k) + \partial/\partial Q(k)] .$$

For interaction with long wavelength, low-energy phonons the deformation potential approximation, frequently in use for covalent compounds with spherical energy surfaces $E = E(k)$, leads to the form [45,88]

$$H' = i\left(\frac{\partial E(k = 0)}{\partial \Omega}\right) \sqrt{\tfrac{1}{2} NM\Omega} \sum_{kq} \sqrt{\omega_k} \, |k| \, (a_k - a_k^*) b_{k+q}^* \, b_q$$

Ω being the volume of the crystal and b, b* electronic operation. The phonons are of the acoustic type.

For ionic crystals and localised electrons (in deep, narrow or shallow extended states) Duke and Mahan [45] gave expressions for additional types of coupling. (Diagonal terms in the electronic states were shown; from these one can derive the off-diagonal terms which will also be used by us).

$$\langle e|H'|e\rangle = V_{kj} M_{ee}(k) \, a_{kj} \tag{2.10}$$
$$+ \text{hermitean conjugate}$$

127

where, for Fröhlich-coupling to longitudinal optic (LO) phonons,

$$V_{k,LO} = \frac{4\pi e}{k} \left[\frac{\hbar\omega_{LO}}{8\pi\Omega} \left(\frac{1}{\epsilon_\infty} - \frac{1}{\epsilon_o} \right) \right]^{1/2}$$

and for piezoelectric coupling to acoustic (A) branches

$$V_{k,A} = \hbar v_s \left[\frac{2\pi^2 C_o^2 e^2 C_{pe}^2}{k \epsilon_o^2 \hbar v_s^3} \frac{C_{pe}^2}{NM} \right]^{1/2} \frac{k^2}{k^2 + k_\Delta^2} .$$

In these expressions ϵ_o and ϵ_∞ are the static and high-frequency dielectric constants of the host, v_s is a (mean) velocity of sound therein, C_{pe} is the piezoelectric coupling constant, C_o a numerical factor that depends on the crystal structure and k_D the Debye inverse screening length.

The matrix elements M_{ee} are unity (apart from phase) for a narrow electronic impurity state. For an s-type hydrogenic, shallow donor state with Bohr-radius a_B $M_{ee}(k) \propto [1 + (k\, a_{B/2})^2]^{-2}$. Thus narrow states couple to short wavelength phonons and broad states to those of long-wavelength. In a general way phonons having wavelengths of the order of the size of the electronic orbits or longer will be most strongly coupled (Toyozawa [152]). In this review article criteria are presented for the appearance in optical spectra of vibrational fine structure, in terms of the phonon-electron interaction.)

2.4.2 Crystal Field Ions

Van Vleck [156a] wrote down the linear interaction for an impurity electron whose potential arises from a high symmetry field (the crystal field). He then made analysis of the coefficients of the coupling, supposing that this arises from the six nearest neighbours around an octahedrally coordinated impurity. Let us exemplify his result for a situation to be used later (Sect. 3.1): linear coupling due to vibrations of E-symmetry (components θ and ϵ) at the impurity, having matrix elements M_{ee}, within an electronic orbital doublet E coming from d-electrons. The result can be expressed in terms of the coefficients V_{kj} in (2.10) and the polarisations introduced in (2.2)

$$e_\alpha(\kappa | kj) \equiv e_\alpha$$

by a formula, valid for long-wavelength acoustical phonons (Van Vleck [156b])

$$V_{kj} = V_E \left(\frac{\hbar\omega}{3 M\Omega\, v_s^2 k^2} \right)^{1/2} \begin{cases} 2e_z k_z - e_x k_x - e_y k_y & (\theta\text{-component}) \quad (2.11) \\ e_x k_x - e_y k_y & (\epsilon\text{-component}) \quad (2.12) \end{cases}$$

where

$$V_E = \frac{3\sqrt{3}}{7} \frac{Z e^2 \langle r^2 \rangle}{a^4} - \frac{25\sqrt{3}}{63} \frac{Z e^2 \langle r^4 \rangle}{a^6} \qquad (2.13)$$

Z being the charge on each neighbouring ion, a the distance from the centre and $\langle r^n \rangle$ an electronic radial average.

More generally, supposing that the effect of the vibrational displacements on the impurity electron is equivalent to that of a small amplitude macroscopic strain one obtains for very long wavelengths

$$V_{kj} \propto k^{\frac{1}{2}}$$

Some care must be exercised if one wishes to include changes in the elastic constant near the impurity from that of the host lattice [18] or effects of the relaxation of the lattice near the impurity [39].

2.4.3 Anticipated Developments in Electron-Vibration Interaction

One would expect that more reliable results will be available before long. These would incorporate effects of chemical bonding (covalency), polarisability and co-operative, many-electron behavior. The amount of sophistication that one can justi-fiably hope for in the estimation of electron-phonon coupling is that which is nowa-days current in the models for calculating vibrational spectra of pure crystals [68]. There improvements in the models from the rigid-ion-like to the deformation dipole model [85], to the shell-model [29,30,86] to the breathing shell-model [114] and lately to the double shell-model [165] have greatly increased the confidence in the quantitative interpretation of data from neutron scattering (e.g. [43,125]) and other techniques [79].

Significant advances have occurred of late in the computation of electron-vibra-tion interactions in molecules. These are based on more or less sophisticated versions of the LCAO-method. A brief review of these works up to about 1978 is now avail-able [50c] (Chap. 8).

3.1 The Formalism of Linear Interaction

The change of potential energy due to the presence of the impurity

$$\delta V = \delta V(r_i, \{l\kappa\}) \tag{3.1}$$

is a function of the coordinates r_i of the impurity electrons and of the positions of all the atoms in the lattice symbolised by $\{l\kappa\}$, as before. In (2.6) we introduced $\overline{\delta V}$ the electronic average of δV. We shall subsequently expand δV in small (normal-mode) displacements, but before that we wish to exhibit the formal dependence of $\overline{\delta V}$ on "the suitably chosen set of electronic states on the impurity". This was briefly mentioned in Sect. 2.3; here we shall make the treatment more precise extending also the formalism of Thomas [150] to electronic states that are non-degenerate.

Let the electronic kets of the set be $|e\rangle$ ($e = 1, \ldots, g$) and their weight factors p_e. Then within this set the operator equivalent expression for δV is

$$\delta V = \sum_{ee'} \delta V_{ee'} |e\rangle\langle e'| \tag{3.2}$$

R. Englman

where

$$\delta V_{ee'} \equiv \langle e | \delta V | e' \rangle \tag{3.3}$$

To provide a consistent formulation of the impurity electron-vibration inter-action we exclude the purely vibrational part in δV, which formed the subject of Sect. 2.3. We thus define the electron-vibration interaction by

$$\delta V_{ev} = \delta V - (\sum_{e'} \delta V_{e'e'} \, p_{e'}) \sum_{e} |e\rangle\langle e|$$

$$= \sum_{e} [V_{ee} - (\sum_{e'} \delta V_{e'e'} \, p_{e'})] \, |e\rangle\langle e| + \sum_{e \neq e'} \delta V_{ee'} \, |e\rangle\langle e'| \tag{3.4}$$

The weighted average of δV_{ev}, namely

$$\sum_{e} (\delta V_{ev})_{ee} \, p_e$$

is zero (in agreement with the redefinition of the normal mode origins in Sect. 2.3). We can therefore write

$$\delta V_{ev} = \sum_{\nu} \delta V_{\nu} \, \sigma_{\nu} \tag{3.5}$$

where σ_{ν} are matrices in the function-space of the "suitably chose set", such that the weighted average of each is zero.

In the cases that the set comprises degenerate states only, σ are clearly traceless matrices. They are familiar from the formalism of the Jahn-Teller effect [50a,66]. Thus for E ⊗ ϵ, a doubly degenerate impurity state interacting with a doubly degenerate mode, we have for the σ-matrices (with $\nu = \theta, \epsilon$ denoting the components of the doubly degenerate representation)

$$\left. \begin{array}{l} \sigma_{\theta} = \begin{pmatrix} -1 & 0 \\ 0 & 1 \end{pmatrix} \\[1em] \sigma_{\epsilon} = \begin{pmatrix} 0 & 1 \\ 1 & 0 \end{pmatrix} \end{array} \right\} \tag{3.6}$$

where the matrices are defined in the $|\theta\rangle, |\epsilon\rangle$ impurity electron space.

The coefficients δV_{ν} originate from the matrix elements (over the impurity electronic states) of the electron-lattice potential function $\delta V \equiv \delta V(r_i, \{l\kappa\})$ introduced in (2.5). Concretely for the E ⊗ ϵ example

$$\delta V_{\theta} = - P_{\theta} \, \langle \theta | \delta V | \theta \rangle \tag{3.7}$$

$$\delta V_{\epsilon} = P_{\epsilon} \langle \theta | \delta V | \epsilon \rangle \tag{3.8}$$

130

where P_ν are symmetry operators (acting on the lattice points $l\kappa$) such that they project out θ and ϵ-representations from the potential.

The expansion of δV in the displacements as in (2.5) then leads (formally) to an expansion of δV_ν in the first and higher powers of u. Such an expansion is a useful first step in the theory, expecially if the potential of the electron is taken as a sum of partial potentials, each due to a different atom in the lattice. Thus for a rocksalt type lattice in which the impurity is a substitutional d-ion and where the potential at the impurity electron is the sum of six Coulombic potentials from the nearest neighbours (numbered as in Fig. 7) one finds

$$\delta V_\theta = \mathscr{V}_E [2 u_z(3) - 2 u_z(6) - u_x(1) - u_x(4) - u_y(2) - u_y(5)]/\sqrt{12} \qquad (3.9)$$

$$\delta V_\epsilon = \mathscr{V}_E [u_x(1) + u_x(4) - u_y(2) - u_y(5)]/2 \qquad (3.10)$$

where $\mathscr{V}_E$ was given in (2.13).

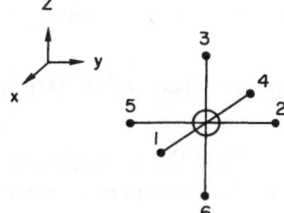

Fig. 7. Numbering of ions in octahedral coordination around the impurity (open circle)

For the dynamic properties of the impurity system the formalism of (3.7)–(3.10) is not convenient, since the total Hamiltonian arises from (2.3), (2.4) as well as (2.5) and in the former two the dynamical variables are the normal mode coordinates of the lattice and their canonical conjugates. When the coefficients δV_ν are written out in terms of these variables, their formal expression up to the second order in $Q_j(k)$ is

$$\delta V_\nu = \sum_{jk} B^\nu_{jk} \, Q_j(k) + \sum_{jj'kk'} C^\nu_{jkj'k'} \, Q_j(k) \, Q_{j'}(k') \, . \qquad (3.11)$$

We recall that the amplitudes $Q_j(k)$ are characterised by the branch-index j and the wavevector k in the lattice. We shall now rewrite the Hamiltonian in terms of normal modes that are adapted to the site of the impurity. These are the analogues of incoming or outgoing spherical waves of scattering theory, whereas $Q_i(k)$ are like running-waves. Both types are eigen-(normal-) modes of the pure lattice Hamiltonian. The precise definition of the site-symmetry adapted normal modes of vibration (SSANMV [51a]) requires some care, but if the reader has no patience with this, he can go over directly to equation (3.18) which represents the starting point for the dynamics of the impurity containing lattice (exclusive of quadratic or higher order terms in the impurity coupling). An alternative pathway leading to SSANMV, originally due to Stevens [143] will be indicated in a later section.

3.2 Symmetry Adapted Coordinates (SSANMV)

It is our purpose to construct a new set of normal displacements

$$q_{\Gamma k^*jn\gamma} \equiv q_{\Gamma i\gamma} \qquad (i = k^*jn) \tag{3.12}$$

whose indexes are explained as follows:
j is the branch index of the mode, as before, Γ is an irreducible representation of the point group at the impurity site, γ is a component of the representation and $n(= 1, 2, ...,)$ labels different SSANMV belonging to the same representation, provided that there are several If not, then $n = 1$ or n is omitted. k^* represents the star of a point k in the reciprocal lattice, namely that set of points which go into each other under all symmetry operations of the point group of the symmetry site. In a cubic lattice, if k is a general point in reciprocal space (that is, one that does not occupy any symmetry element of the group) a star k* consists of 48 points in k-space (48 being the number of elements of this cubic group).

The dimensionless SSANMV amplitudes are projected out of the running mode coordinates $Q_j(k)$ through

$$q_{\Gamma k^*jn} = (\omega_{\Gamma k^*j}/\hbar)^{\frac{1}{2}} (D_\Gamma/gc)^{\frac{1}{2}} \sum_S a_S^{(n)} P_{\Gamma\gamma}(S) Q_j(k) \tag{3.13}$$

Here D_Γ is the dimension of the representation Γ, g the order of the point group, $P_{\Gamma\gamma}(S)$ is an operator corresponding to the symmetry-operation S of the point group that projects out the representation from the operand $[Q_j(k)]$.

The numbers c, n and $a_S^{(n)}$ are connected through the representation matrix $G^\Gamma_{\nu\mu}(S)$ and the following equations

$$\sum_S a_S^{(n)} G^\Gamma_{\nu\nu}(S) = \delta_{n1}$$

$$\sum_{SS'} a_S^{(n)} a_{S'}^{(n')} G^\Gamma_{\nu\mu}(S'S^{-1}) = c\, \delta_{nn'} \tag{3.14}$$

The preceding set of Eqs. (3.13)–(3.14) define a systematic procedure to derive all SSANMV. The impurity electron-vibration Hamiltonian can be written in terms of these as

$$H_{ev} = \frac{1}{2} \sum_{\Gamma k^*jn} \hbar\, \omega_{\Gamma k^*j} [p^2_{\Gamma k^*jn\gamma} + q^2_{\Gamma k^*jn\gamma} + 2k_{\Gamma k^*jn}\, q_{\Gamma k^*jn}\, \sigma_{\Gamma\gamma}] \tag{3.15}$$

where $\sigma_{\Gamma\gamma}$ are the traceless matrices (or combinations thereof) introduced in (3.5), transforming as the $\Gamma\gamma$ irreducible representation of the point group. In fact there may be more than one matrix having the same transformation, then one should write $\sigma^{(l)}_{\Gamma\gamma}(l = 1, 2, ...,)$[51b]. Since the frequencies ω do not depend on n, we see from the last equation that out of the D_Γ codegenerate SSANMV we can form one linear combination that alone is subject to the electron-vibration coupling. This combination is (for each $\Gamma k^*j\gamma$)

$$q^I_{\Gamma k^*j\gamma} = (k^I_{\Gamma k^*j})^{-1} \sum_{n=1}^{D_\Gamma} k_{\Gamma k^*jn} \, q_{\Gamma k^*jn\gamma} \tag{3.16}$$

where

$$k^I_{\Gamma k^*j} = (\sum_{n=1}^{D_\Gamma} k^2_{\Gamma k^*jn})^{\frac{1}{2}} \tag{3.17}$$

In the instance of MgO:Fe[27], which in its ground state is a $T \otimes \epsilon$-type Jahn-Teller impurity within cubic symmetry, a single coupled mode equivalent to (3.16) was used[67]. From our construction it is evident that there will also be $D_\Gamma - 1$ linear combinations $q^\tau_{\Gamma k^*j\gamma}$ ($\tau = 2, \ldots, D_\Gamma$) arising from $q_{\Gamma k^*jn\gamma}$ and orthogonal to $q^I_{\Gamma k^*j\gamma}$ that are not coupled to the impurity electron.

We can now rewrite the electron-vibration Hamiltonian in the more economic form

$$H_{ev} = \frac{1}{2} \sum_{\Gamma k^*j\gamma} \hbar\omega_{\Gamma k^*j}[(p^I_{\Gamma k^*j\gamma})^2 + (q^I_{\Gamma k^*j\gamma})^2 + 2k^I_{\Gamma k^*j} \, q^I_{\Gamma k^*j} \, \sigma_{\Gamma\gamma}]$$

$$+ \frac{1}{2} \sum_{\Gamma k^*j\gamma} \sum_{\tau=2}^{D_\Gamma} \hbar\omega_{\Gamma k^*j}[(p^\tau_{\Gamma k^*j\gamma})^2 + (q^\tau_{\Gamma k^*j\gamma})^2] \tag{3.18}$$

In later Sect. 4.2, 4.4.2, we shall further simplify the interaction term in this Hamiltonian, with the intent to reduce the number of interacting modes (of which there are of the order N in (3.18)) to a small number on the order of unity. (3.18) is still useful when the electron-vibration coupling can be applied perturbationally. Examples of this are given in the next section. In the other cases that the coupling has to be taken to high order, (3.18) is inconvenient (or impossible) to work with. With these cases in mind several alternatives to (3.18) have been derived and these will be described subsequently.

4 Applications of the Electron-Vibration Coupling Using Different Descriptions

4.1 Phonon Description

4.1.1 The Spin-Phonon Relaxation in a Paramagnetic System

We take for our example a system of two-spin-levels separated by a Zeeman-splitting of $\hbar\omega_0$. In the matrix description of these two levels, the off-diagonal spin-lattice coupling is given by (3.15) or (3.16) with the relevant matrix being

$$\sigma_{\Gamma\gamma} = \sigma_x = \begin{pmatrix} 0 & 1 \\ 1 & 0 \end{pmatrix} \tag{4.1}$$

while the physical origin of the coupling constants k is the electron-phonon and spin-orbit coupling mechanisms combined, as discussed in detail by Abragam and Bleaney [3] and by Orbach and Stapleton [119].

In this representation the Zeeman splitting is written

$$H_Z = \hbar\omega_o \, \sigma_z \tag{4.2}$$

which is to be added to the Hamiltonian H_{ev}. In a characteristic ESR situation H_Z is small, and so is the electron-vibrational coupling, in the sense of the sum (integral) of the squares of all k's for all stars (k*) being much smaller than unity.

Suppose now the usual situation in an ESR experiment when the spin system is excited by an rf field to the upper of the two Zeeman levels and let us calculate its rate of decay to the lower level, supposing a one-phonon (direct) relaxation process. Using time-dependent perturbation and taking the average over the initial and sum over the final phonon states, one obtains the following result for the rate of decay

$$W = \pi[\omega_o \, k(\omega_0)]^2 \, [n(\omega_o) + 1] \rho(\omega_o) \tag{4.3}$$

Here ρ is the density of phonons, $n(\omega)$ is the Planck number corresponding to the thermal distribution of phonon excitations and $k(\omega)$ is the spin-phonon coupling constant written as a function of frequency (instead of the wave-vector star k* and the phonon branch index, as previously). Only those phonons with energy equal to the Zeeman energy $\hbar\omega_o$ are of interest in a direct relaxation process. This energy is characteristically 0.1 cm^{-1} and the relevant phonons are of the long-wave acoustic type. Their role is to modulate the crystal field interacting with the electron.

For non-Kramers levels (integral pseudo-spin) $k(\omega_o)$ goes as $\omega_o^{-\frac{1}{2}}$, while for a Kramers doublet (spin states $\pm\frac{1}{2}$) the coupling is a combined effect of strain and of magnetic field and then $k(\omega_o) \propto \omega_o^{\frac{1}{2}}$.

4.1.2 The $E \otimes \sum_i \epsilon_i$ Vibronic System in the Weak Coupling Limit

For a doubly degenerate electronic state E (components $|\theta\rangle$, $|\epsilon\rangle$) coupled to a quasi-infinite set of SSANMV, each transforming as E in the point group of the impurity, the general expression (3.18) takes the form:

$$H_{ev} = \sum_i \left[\sum_{\gamma=\theta,\epsilon} \frac{1}{2} \hbar\omega_i(p_{i\gamma}^2 + q_{i\gamma}^2) + \hbar\omega_i \, k_i(q_{i\theta} \, \sigma_z + q_{i\epsilon} \, \sigma_x) \right] \tag{4.4}$$

in which only the E-transforming modes have been included. σ_z and σ_x appear in (3.6). The eigenstates of H_{ev} can be labelled quite generally with the eigenvalues j of the total angular momentum operator, defined by

$$\hat{j} = j_{vib} + j_{el} = \sum_i (q_{i\theta}\, p_{i\epsilon} - q_{i\epsilon}\, p_{i\theta}) - \frac{1}{2}\, \sigma_y$$

$$\sigma_y = \begin{pmatrix} 0 & i \\ -i & 0 \end{pmatrix}$$

A formal solution to (4.4) is the pair of states

$$\left|+\tfrac{1}{2}\right\rangle = f_0\, |+\rangle + g_1\, |-\rangle$$

$$\left|-\tfrac{1}{2}\right\rangle = \left|+\tfrac{1}{2}\right\rangle^* = f_0^*\, |-\rangle + g_1^*\, |+\rangle \tag{4.5}$$

where $|\pm\rangle = (|\theta\rangle \mp i\,|\epsilon\rangle)/\sqrt{2}$ belong to the eigenvalues $j_{el} = \pm\frac{1}{2}$ while f_0, g_1 and their complex conjugates are functions of the coordinates $q_{i\gamma}$. In the ground vibronic doublets $|j| = \frac{1}{2}$ and the eigenvalues of f_0 and g_1 are $j_{vib} = 0,1$.

In many cases involving degenerate ionic impurities the linear Jahn-Teller coupling is strong ([50b], Appendix IX)

$$K \equiv \sum_i k_i^2 > 1 \tag{4.6}$$

and solutions correct to all orders of K are required. The many phonon description of (4.4) is difficult to handle and further manipulations of the vibrational modes are required to solve the problem. This is the subject of the Sect. 4.3. For the present weak coupling $K \ll 1$ is presupposed and the last term in (4.4) is treated perturbationally (correct essentially to K^2). With this restriction one can represent the lattice motion by perturbed phonons.

The vibrational factors f_0, g_1 and g_{-1} of (4.5) can be expressed in terms of creation operators $a_{i\gamma}^+$ of SSANMV operating on the ground state function of non-interacting phonons; namely

$$\Psi_{gr} = \prod_i \prod_{\gamma=\theta,\epsilon} u_o(q_{i\gamma}),$$

(u_o being the ground state wave-function of a single vibration) and to second order in the coupling strengths k:

$$f_0 = \{1 - \frac{1}{2}\sum_i k_i^2\,[1 - \frac{1}{2}(a_{i\theta}^{+2} + a_{i\epsilon}^{+2})] + \frac{1}{4}\sum_{i,l=i} k_i\, k_l\,[a_{i\theta}^+ a_{l\theta}^+ + a_{i\epsilon}^+ a_{l\epsilon}^+ +$$

$$+ i\left(\frac{\omega_i - \omega_l}{\omega_i + \omega_l}\right)(a_{i\theta}^+ a_{l\epsilon}^+ - a_{i\epsilon}^+ a_{l\theta}^+)]\}\, \Psi_{gr} \tag{4.7}$$

$$g_1 = \left[\frac{1}{\sqrt{2}}\sum_i k_i(a_{i\theta}^+ + i\, a_{i\epsilon}^+)\right]\Psi_{gr} \tag{4.8}$$

The vibronic ground state energies are

$$E = \sum_i \hbar\omega_i(1 - k_i^2)$$

135

An interesting feature of (4.7) is that the imaginary terms in it are proportional to $(\omega_i - \omega_l)$. These vanish only if all modes that are coupled to the impurity electron $(k_i \neq 0 \neq k_l)$ have identical frequencies $(\omega_i \equiv \omega_l)$, or equivalently if there is one interacting mode. [Actually in an Einstein model of the crystal where all frequencies are equal, one can clearly select one interacting mode]. Otherwise, for a spread of frequencies

$$\text{Im } f_0 \neq 0 .$$

For the interpretation of experiments on doubly degenerate ions, further terms need to be added to the Hamiltonian H_{ev}. In general one has to include the effects of strain, of electric field and, if the electrons are not in a spin-singlet state, also of magnetic field and hyperfine interaction. Spin-orbit effects vanish within an E state but they can reappear in higher order (in the spin or through combination with external fields) by coupling to orbital levels outside the doublet. The objectives of this Report do not justify a detailed description of these effects, already summarised by Ham [66] and brought up-to-date by Bates [14], so we consider only some phenomena that arise from the application of external stresses. These result in macroscopic strains whose components at the impurities are:

an isotropic strain, e_A
two components e_θ, e_ϵ of lower symmetry strain field, that transform as the corresponding irreducible representations E at the impurity
additional strain of symmetry other than the above.

The interaction of a strain with an electronic doublet can be conveniently represented by matrices in the electronic function space. Our concern is with the effects of strain on the ground vibronic doublet. This has the form (4.5) where the vibrational functions f_0, g_1 are for weak coupling as in (4.7), (4.8). The strain Hamiltonian can be represented in the space of the electronic functions $\pm >$ by

$$H = G(A_1)\, e_A \begin{bmatrix} 1 & 0 \\ 0 & 1 \end{bmatrix} - G(A_2)\, p \begin{bmatrix} 1 & 0 \\ 0 & -1 \end{bmatrix} - G(E)\, q \begin{bmatrix} 0 & e_\theta + i\, e_\epsilon \\ e_\theta - i\, e_\epsilon & 0 \end{bmatrix} \quad (4.9)$$

where the coefficients $G(\Gamma)$ incorporate the effects of the strain on the electronic states [$G(A_2)$ requires another agent, e.g. a magnetic field in addition to the low symmetry strain] while the reduction factors q and p reflect the presence of the vibronic coupling [66]. In its absence:

$$q = 1 , \quad p = 0 .$$

Following several workers [60,65a,b,108], we can write a relation between the reduction factors in the form

$$2\, q - p = 1 - 2\, \langle \text{Im } f_0 \,|\, \text{Im } f_0 \rangle \quad (4.10)$$

We recall from (4.6) that Im f_0 is non-zero in any genuinely multimode vibronic system; from this we conclude that the left hand side of (4.10) is less than unity. This conclusion holds true for coupling of any strength, though this section has dealt only with weak coupling.

A confirmation of the many-mode coupling model was proposed[1] for Al_2O_3:Ni^{3+} by experimental determination of the left-hand-side in (4.10). As already described in Sect. 1.3, Ni^{3+} has a ^2E ground state in the trigonally symmetric Al_2O_3 and is subject to strong vibronic coupling of the type appearing in (4.4). The reduction factors q and p were derived from thermally detected electric-field-induced ESR data and from acoustic paramagnetic resonance. In the former type of experiments a temperature rise in the cavity of about 0.01 $K°$ is monitored as function of the constant magnetic field, with the rf kept fixed, as is usual in ESR work. Vibronic levels were identified partly through the variation of the ambient temperature and the accompanying observation of changes in the intensities of several peaks.

The work of the Nottingham group led to the value of q = 0.467 not far from 0.5, appropriate to extremely strong Jahn-Teller coupling[66]. Higher order vibronic couplings (e.g. cubic in the vibrational coordinates) were found to be less important than previously supposed (Sturge et al. 1967[146]). When p was experimentally determined, the left-hand-side of (4.10) was found to vary between 0.52 and 0.80[14]. Valuable experimental support, confirmed also by subsequent observations[134], was therefore provided to the multi-mode nature of the interacting vibrations.

4.2 The Molecular Description and its Uses

4.2.1 Survey

The diametrically opposite treatment to the phonon description in the previous sections is to regard the impurity and its immediate neighbours in the solid as an isolated unit, or a molecule. The vibrational motion of the atoms in such a "molecule" is represented by the molecular normal modes, with amplitudes $Q_{\Gamma\gamma}$ (having dimensions as in (2.5) of length x square root of the reduced mass M) and their-momenta-conjugates $P_{\Gamma\gamma}$ (Γ is an irreducible representation of the molecular point group and γ its component). The electron-vibration coupling enters the Hamiltonian as

$$H_{ev} = \frac{1}{2} \sum_{\Gamma\gamma} (P_{\Gamma\gamma}^2 + \omega_\Gamma^2 Q_{\Gamma\gamma}^2) + \sum_{\Gamma\gamma} V_\Gamma M^{-\frac{1}{2}} Q_{\Gamma\gamma} \sigma_{\Gamma\gamma} \qquad (4.11)$$

in which the electronic matrices $\sigma_{\Gamma\gamma}$ are identical to those in the previous section (Eqs. (3.18) or (4.9)). The construction of molecular normal modes from either experimental infra-red and Raman data or force models is a well established practice[109].

Detailed expressions of the coupling strengths V_Γ were given by Bates[14] for regular octahedral (XY_6) and tetrahedral (XY_4) molecules where the ligands Y were represented by point charges and the impurity atom X has several atomic terms arising from d^n (n = 1, ... , 10) configurations. The electronic cloud of X was considered internal to the ligands and covalency was not treated. The formalism of Bates[14] used electronic operators ("T") rather than their matrix representatives (σ).

The quasi-molecular Hamiltonian (4.11) has had an immensely rich past as a model for point impurities in crystals. For reasons of symmetry and also of the wish for simplification only a few modes were normally included in the second sum in (4.11). These modes have been named "interaction-", "cluster-" or "configurational-" modes. Although as we have remarked, the range of application is very wide we have made a very narrow selection of those instances in which there has been significant experimental information on the *character* of the interaction mode.

4.2.2 Configuration Coordinate Description of Optical Transitions

The use of molecular description for electronic transitions in the optical range dates back to the 1950's[123,166]. This case differs from those considered earlier in this report in that one has characteristically two (or more) electronic multiplets which are energetically spaced far apart, in the sense of the "energy gap" ΔE (= the separation between two electronic states) being very much larger than the vibrational energies $\hbar\omega$ appearing in, say, (4.4).

Let Γ_u and Γ_l be the representations of the two (upper and lower) electronic states participating in the optical transition. Then the modes which will participate in the electron-vibration interaction will have representations that are included in the symmetric square of $\Gamma_u + \Gamma_l$. Thus, for the transition $A_{1g} \to T_{1u}$ (the C band) in KCl:Tl$^+$ type phosphors, modes of representations

$$[A_{1g} + T_{1u}]^2 = 2 A_{1g} + E_g + T_{2g} + T_{1u} \tag{4.12}$$

will be coupled to the electrons. Moreover, each mode will in fact occur only once (although A_{1g} appears in (4.12) twice), since in our discussion in Sect. 2.3, (2.7), we have used the coupling to the unit matrix (belonging to A_{1g}) to redefine the origin of the normal coordinate. In fact, there are cases in which a representation may occur in the coupling more than once[51b], but in the present report, whose purpose is erstwhile pedagogic, we shall avoid complicating situations.

In writing out explicitly the electron-vibration interaction term appropriate to an $A_{1g} \to T_{1u}$ optical transition, we first of all neglect T_{1u}-type vibrations. These are non-diagonal in the initial and final states, whose separation is large compared to the vibronic interaction strength. The electronic matrices $\sigma_{\Gamma\gamma}$ operating on the 4-vector $(A_{1g}, T_{1ux}, T_{1uy}, T_{1uz})$ are as follows

$$\sigma_{A_{1g}} = \frac{1}{2} \begin{pmatrix} 1 & 0 & 0 & 0 \\ 0 & -1 & 0 & 0 \\ 0 & 0 & -1 & 0 \\ 0 & 0 & 0 & -1 \end{pmatrix} \tag{4.13}$$

$$\sigma_{E_g,\theta} = \frac{1}{\sqrt{6}} \begin{pmatrix} 0 & 0 & 0 & 0 \\ 0 & -1 & 0 & 0 \\ 0 & 0 & -1 & 0 \\ 0 & 0 & 0 & 2 \end{pmatrix}$$

$$\sigma_{E_g,\epsilon} = \frac{1}{\sqrt{2}} \begin{pmatrix} 0 & 0 & 0 & 0 \\ 0 & 1 & 0 & 0 \\ 0 & 0 & -1 & 0 \\ 0 & 0 & 0 & 0 \end{pmatrix}$$

$$\sigma_{T_{2g},\epsilon} = \begin{pmatrix} 0 & 0 & 0 & 0 \\ 0 & 0 & 1 & 0 \\ 0 & 1 & 0 & 1 \\ 0 & 0 & 1 & 0 \end{pmatrix}, \qquad \text{etc.}$$

The normalizations of these conform to those in the literature [15], excepting $\sigma_{A_{1g}}$ whose form as well as normalization are to some extent arbitrary. This is due to there being two independent matrices belonging to this representation, another one being

$$\sigma_{A_{1g}}^{(2)} = \begin{pmatrix} 3 & 0 & 0 & 0 \\ 0 & 1 & 0 & 0 \\ 0 & 0 & 1 & 0 \\ 0 & 0 & 0 & 1 \end{pmatrix}$$

The choice in (4.13) is convenient for discussing optical transitions due to light having a definite polarisation, say along z, since with the matrix as in (4.13) the equilibrium positions of the $Q_{A_{1g}}$ coordinate in the initial (A_{1g}) and final (the z-component of T_{1u})-states are displaced to the left and right by equal amounts. This will be apparent in the next but one equation giving the electron-vibration interaction. The relevant total Hamiltonian contains also an electronic term H_{el}

$$H = H_{ev} + H_{el}$$

where

$$H_{ev} = \sum_{\Gamma\gamma} \frac{1}{2} (P_{\Gamma\gamma}^2 + \omega_\Gamma^2 \, Q_{\Gamma\gamma}^2) + \sum_{\Gamma\gamma = A_{1g}, E_g, T_{2g}} V_\Gamma \, M_\Gamma^{-\frac{1}{2}} \, \sigma_{\Gamma\gamma} \qquad (4.14)$$

$$H_{el} = - \, \Delta E \, \sigma_{A_{1g}}$$

In (4.14) the coefficients V_Γ depend on the nature of (or model for) the molecule. $V_{A_{1g}}$ is defined in terms of the $Q_{A_{1g}}$-derivative of the electronic expectation value of the potential, namely

$$\langle e \, | V | \, e \rangle \equiv V_{ee} \qquad\qquad (e = A_{1g}, T_{1ux}, \ldots) \ldots$$

as follows:

$$V_{A_{1g}} = \frac{1}{6} (3 \, V_{A_{1g},A_{1g}} - \sum_{\xi=x,y,z} V_{T_{1u\xi}}) \qquad (4.15)$$

i.e. the weighting factors p_e, introduced in (2.6), are $\frac{1}{2}, \frac{1}{6}, \frac{1}{6}, \frac{1}{6}$.

4.2.3 Experimental Determination of the Parameters

Following Seitz's [136] inception of configuration coordinates the molecular description was successfully used to interpret optical spectra of impurities [59,127,144]. Frequently it is sufficient to utilise a single interacting molecular mode only [in (4.11)] or the data does not warrant a more profound treatment. Several F-centres are like this [71]. Then one has but two theoretical parameters the mode frequency ω and the linear electron-vibrational coupling constant $V/M^{\frac{1}{2}}$. Thus the question inevitably arises: what is the experimental backing of the model and what is its predictive value? Normally, from absorption spectra one derives three quantities: the Stokes-shift, the width of the band and the ratio of the zero-phonon-line and band intensities, and assigns the parameters $\hbar\omega$ and S (the Huang-Rhys number;[75]) this is related to the parameters in (4.11) by $S = V^2/2\,\hbar\,\omega^3\,M$). One can further check for consistency by computation of higher moments, utilization of magnetic circular dichroism and the temperature variation in the experimental quantities [144] (Chap. 10).

Particular interest attaches to cases where a series of centres of the same type exists. Then there is some chance for systematisation and the capacity to predict. The famous F-band data by Dawson and Pooley (1969) (also in Fowler [59] and Henderson [71]) for a number of alkali halides are an example in hand. For these the mode frequencies follow a pattern (i.e. correlate with the optical frequencies of the host crystal) but no compelling rationalisation of the coupling-strength parameter S is known. In another series: heavy ion phosphors of the type M^+: alkali halide, a systematic variation of the band properties is evident as, e.g., M is varied by descending along a column in the periodic table [35] since the spin-orbit coupling increases steadily, but once more the electron-vibration interaction defies systemisation. A partial reason for this is that in these phosphors several modes of diverse symmetry are coupled.

A successful instance of empirical correlation was found for the electron-vibrational coupling, between ^2E_g and $^4A_{2g}$-states on Mn^{4+}, entering substitutionally for M in the octahedral sites of Cs_2MF_6 (M = Si, Ge, Ti, Sn, Zr) and M_2SiF_6 (M = K, Rb, Cs) [121]. The Huang-Rhys number S appears in the intensity of the n'th vibronic sideband of a progression as $\exp(-S)S^n/n!$. From comparison with his emission data at $80\,°K$ Paulusz found a quadratic dependence of S on the estimated Mn–F distance for both A_{1g} and E_g vibrational modes in the former series of hosts and a virtual constancy for the latter series.

Earlier Blasse (1969) [20] argued that upon excitation of an impurity cation the molecular mode coordinate presses outwards due to the oxidation of the ion by charge transfer. He correlated the coordinate change (a quantity proportional to V) with the quenching temperature of luminescence, which he determined from experiments with $MeWO_4$ (Me is an alkaline earth) and rare-earth ions in YBO_3 and $CaBPO_5$ type compounds. Extending this work to impurity ions with excess charge, Blasse (1974) [20] found in a series of impurity centres (of which the following are prototypes: $YPO_4:W^{6+}$, $CaSO_4:Ti^{4+}$, $YPO_4:Ti^{4+}$, $Y_2O_3:Pb^{2+}$, etc.), a quenching of luminescence on condition that the extent of the cluster increases upon excitation. The charge-compensating defect played no role in this criterion. This circumstance could find its cause either in a photochemical reaction of the impurity centre or in the nature of the interacting mode.

Radlinski observed [126] the absorptions $^4A_2(^4P) \rightarrow {}^4T_2(^4F)$ on Co^{2+} in several zinc-blende type hosts and determined the electron-phonon coupling strength in each of them. These were found [65e] to agree quite well with a model (based on crystal field theory and a Debye phonon spectrum with limiting frequency ω_D) which gave for the Huang-Rhys number the relationship $S \propto \omega_D^{-3} a^{-12}$ (a is the lattice constant).

4.3 Analysis of Vibrations of Lattice Stars

This approach was originally proposed by Stevens [143] and then further developed by Steggles [142]. It was summarized in a review by Bates [14]. From a logical point of view the method can be regarded as an extension of the molecular description (Sect. 4.2) to larger and larger "molecules" formed by adding further coordination poly-hedra around the impurity in the crystal. In the NaCl structure the first coordination polyhedron constitutes the XY_6 "molecule" considered in the previous section. More distant polyhedra of equivalent atoms increase the size of the structure, maintaining at the same time cubic symmetry. It is expected that by adding one or more poly-hedra to the nearest neighbours the molecular description of Sect. 4.2 would be im-proved in a quantitative way, as in the cluster models of Messmer and Watkins [103], without changing it qualitatively. In actuality Stevens [143] and his coworkers devel-oped their approach in a different direction, aiming at the inclusion of all polyhedra, near or far. In a cubic structure there will be 48 equivalent atoms that form a coordi-nation polyhedron and which constitute a *star* of atoms, similar to the stars k* that

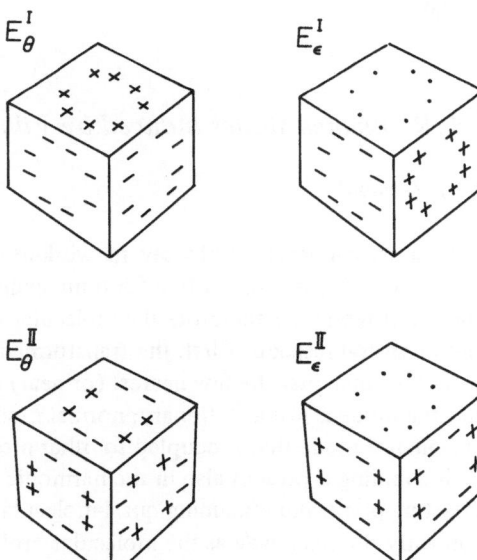

Fig. 8. Displacement pattern of a star of 48 (general) lattice points. The two components θ and ϵ of E-type modes arising from radial displacements of a star are shown with + representing out-ward displacement, − inward and a dot no displacement. The E^I-modes are coupled to electronic motion at the centre, the E^{II}-modes are uncoupled. After Stevens [143] and Bates [14]

141

lie in reciprocal space (Sect. 3.1). This is true provided the atoms do not lie on a symmetry element of the structure; if they do, the number of atoms in the star is less. Figure 8 shows normal displacements of a (general) star $l*$ arising from all lattice points $l = (l_x, l_y, l_z)$ which go into each other under the operations of the cubic group. Out of the 3 x 48 degrees of freedom of a general star the following modes of vibration can be formed

$3 A_{1g}, 3 A_{2g}, 6 E_g, 9 T_{1g}, 9 T_{2g}$ even-modes

$3 A_{1u}, 3 A_{2u}, 6 E_u, 9 T_{1u}, 9 T_{2u}$ odd-modes

By group theory, all modes in a lattice belonging to the same species in a lattice interact (while those of different species are uncoupled) so that if a mode on a neighbouring star will be coupled to the impurity-electron, modes of the same species on more distant stars will be coupled, too. In the works quoted a set of coupled equations of motion was derived for the modes, using relatively simple models for the couplings between atoms. Disturbances that originate near the impurity decrease in amplitude as they spread outwards: their behaviour as function of the distance of the star from the impurity will be discussed in Sect. 4.4.3.

To bring the Hamiltonian of the star-displacements to a canonical form, a transformation was applied [143]. The resulting eigenmodes appear to be equivalent to the SSANMV introduced in (3.13), though the algebra necessary to demonstrate this rigorously is arduous.

For a diamond lattice containing a nitrogen impurity or a vacancy a large but finite cluster model (of up to 70 atoms) was employed by Watkins and Messmer [162] to calculate the distortional modes due to electron-vibrational coupling. From their computed data it appears that the distortion takes place along several normal modes of the cluster. This may be also the consequence of the anharmonic nature of the model.

4.4 Resonance Modes Induced by Vibronic Interaction (Cluster Modes)

4.4.1 General

In an approach developed by several workers [65b–d,51a,57] the many-mode Hamiltonian of (3.15) is brought to a form involving essentially one or a few set of modes. The result resembles therefore the molecular description of Sect. 4.2. It differs from that in several respects. First, the transformed Hamiltonian describes interaction to ions other than just the few nearest (or near) neighbours. Second, it clearly shows that the mode is not a stationary, normal mode but rather, in the sense of Sect. 1.4, a resonance mode that is coupled to other modes by what is called a residual coupling. Such coupling is present also in the harmonic approximation. When it is neglected, as a starting point, the remaining quasi-molecular or cluster Hamiltonian can be tackled in exactly the same way as the molecular problem. This of course is the attractive feature of the cluster-mode approach.

4.4.2 The Transformed Mode

New sets of coordinates of $\tilde{q}_{\Gamma\gamma h}$ are introduced by applying a further orthogonal transformation A_{hi}^Γ on the SSANMV defined in $(3 \cdot 16)$ (the two indices $k*j$ there are now abbreviated by i)

$$\tilde{q}_{\Gamma\gamma h} = \sum_i A_{hi}^\Gamma q_{\Gamma\gamma i}^I .$$ (4.16)

The purpose of this transformation is to bring the Hamiltonian (3.18) to a form consisting of three Hamiltonians: a purely harmonic or phonon-like part, a quasi-molecular (QM) and a third perturbational term H' (the residual coupling)

$$H = H_{\text{phonon-like}} + H_{\text{QM}} + H'$$ (4.17)

$$H_{\text{phonon-like}} = \frac{1}{2} \sum_{\Gamma\gamma} \sum_{h=1} \hbar \Omega_{\Gamma h} (\tilde{p}_{\Gamma\gamma h}^2 + \tilde{q}_{\Gamma\gamma h}^2)$$ (4.18)

$$H_{\text{QM}} = \sum_{\Gamma\gamma} \hbar \Omega_\Gamma (\frac{1}{2} \tilde{p}_{\Gamma\gamma 1}^2 + \frac{1}{2} \tilde{q}_{\Gamma\gamma 1}^2 + \tilde{q}_{\Gamma\gamma 1} K_\Gamma \sigma_{\Gamma\gamma})$$ (4.19)

$$H' = \sum_{\Gamma\gamma} \sum_h c_{\Gamma h} \tilde{q}_{\Gamma\gamma h} K_\Gamma \sigma_{\Gamma\gamma} + \sum_{\Gamma\gamma} \sum_{\substack{hh' \\ h>h'}} d_{\Gamma hh'} (\tilde{p}_{\Gamma\gamma h} \tilde{p}_{\Gamma\gamma h'} + \tilde{q}_{\Gamma\gamma h} \tilde{q}_{\Gamma\gamma h'})$$ (4.20)

The parameters in (4.17)–(4.20) are:
The resonance frequencies

$$\Omega_\Gamma = \sum_h (A_{1h}^\Gamma)^2 \omega_{\Gamma h}$$ (4.21)

The effective coupling strengths of the resonance modes:

$$K_\Gamma = \Omega_\Gamma^{-1} \sum_h A_{1h}^\Gamma k_{\Gamma h} \omega_{\Gamma h}$$ (4.22)

The frequencies of the phonon-like modes

$$\Omega_{\Gamma k} = \sum_h (A_{kh}^\Gamma)^2 \omega_{\Gamma h} \qquad (k \neq 1) .$$ (4.23)

The coefficients in the residual coupling H'

$$c_{\Gamma k} = K_\Gamma^{-1} \sum_h A_{kh}^\Gamma k_{\Gamma h} \hbar \omega_{\Gamma h}$$

$$d_{\Gamma kk'} = \sum_h A_{kh}^\Gamma A_{k'h}^\Gamma \hbar \omega_{\Gamma h} .$$

It is to be emphasized that (4.17) is exactly equivalent to the original Hamiltonian (3.18). The transformed-form is advantageous provided that the residual coupling is small compared to the effective Hamiltonian

$$H_{eff} \equiv H - H' .$$

If this is the case, we can say that the main effect of the impurity electron is felt by the resonance mode, h = 1, given explicitly by

$$\tilde{q}_{\Gamma\gamma 1} = \sum_i A^{\Gamma}_{1i} \, q^I_{\Gamma\gamma i} . \tag{4.24}$$

The other modes $h \neq 1$ in (4.16) feel only perturbational coupling through H'. The importance of H' can be estimated through the sum rules [116]

$$\sum_k d^2_{\Gamma k 1} = \hbar^2 (\langle \omega^2 \rangle_{\Gamma} - \langle \omega \rangle^2_{\Gamma}) \equiv \hbar^2 \langle \Delta \omega^2 \rangle_{\Gamma} \tag{4.25}$$

and

$$\sum_k c^2_{\Gamma k} \propto \sum_k d^2_{\Gamma k 1} .$$

The averages in (4.25) are defined through

$$\langle \omega^n \rangle_{\Gamma} \equiv \sum_h (A^{\Gamma}_{1h})^2 \, \omega^n_h . \tag{4.26}$$

For the residual coupling H' to be negligible we require that the weighted frequency deviation in (4.25) be small, or more precisely that

$$\langle \Delta \omega^2 \rangle_{\Gamma} \equiv \langle \omega^2 \rangle_{\Gamma} - \langle \omega \rangle^2_{\Gamma} \ll \langle \omega \rangle^2_{\Gamma} . \tag{4.27}$$

The physical meaning of this criterion is that the resonance mode $(\tilde{q}_{\Gamma\gamma 1})$ be made up of modes in only a narrow frequency range. This will also ensure a long lifetime, due to H', for the resonance mode.

To discuss the finiteness of this lifetime in some more detail we note from (4.21) that the location of the resonant eigenfrequency Ω_{Γ} is necessarily below the maximal frequency of the pure lattice. In effect, it can be inside a phonon-band or else lie in a forbidden energy region. We shall see shortly that phonon-bands of the pure lattice and of the phonon-like frequencies in (4.23) nearly coincide, so that one does not have to differentiate between the two. However, consideration must be given to the existence of several phonon bands, superimposed on several vibronic levels which arise from the degeneracy of the electronic state (Fig. 9).

If the resonance frequency is inside a phonon band, H' will in general admix bands-modes in the resonance mode, so that in effect the inverse lifetime of the mode

Fig. 9. Structures of vibrational spectra in impure crystals. **(a)** represents impurities whose electronic states (e, e', . . .) are uncoupled from the vibrations of the crystal. The phonon-spectra are superimposed on the eletronic levels (shaded area). Resonance (ω_R) or localised (ω_l) levels may appear. **(b)** If the impurity states are coupled to local vibrations, vibronic levels (v, v', . . .) appear whose spacings are generally much closer than for the electronic levels in **(a)**. The superimposed phonon structures will fill in the energy range

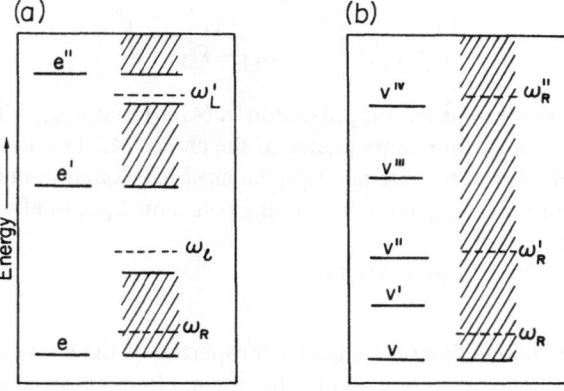

will be proportional to H'^2. By (4.25) this will vary as the square phonon deviation $\langle \Delta\omega^2 \rangle$. Even if the resonance frequency falls in a forbidden band-gap the mode will live finitely, but this effect will only arise in higher order in $(H')^2$ and thus will be reduced in importance. In conclusion, under all circumstances the coupled mode will not be stationary, though it may show a sharp resonance.

Recalling (4.16) and (4.24) we see that the resonance mode depends only on the elements in the first row of the transformation matrices A. These elements can be derived using variational procedures, which are appropriate for ground state properties. When other vibronic states (or a combination of these) are of interest, as e.g. in optical absorption, different procedures might be employed. Such methods are currently being developed. (J.R. Fletcher, M.C.M. O'Brien, private communications, 1979).

In a first treatment of this type of problems, O'Brien[116] varied the Jahn-Teller stabilisation energy of an $E \otimes \sum_i \epsilon_i$ situation, Sect. 3.1,

$$E_{JT} = \frac{1}{2} \hbar \Omega_\Gamma K_\Gamma^2 \qquad \Gamma = E \tag{4.28}$$

with K_Γ given in (4.22). The variation led to the result for the first row in the transformation matrix:

$$A^\Gamma_{1h} = \omega_{\Gamma h} k_{\Gamma h} f(\omega_{\Gamma h}) \tag{4.29}$$

where $f(\omega_{\Gamma h})$ had the particularly simple form

$$f(\omega_{\Gamma h}) = (\kappa_\Gamma \omega_{\Gamma h})^{-1} . \tag{4.30}$$

The elements of the remaining rows are as yet arbitrary, but if one chooses

$$A^{\Gamma}_{hi} = \frac{k_{\Gamma i}}{(\omega_{\Gamma i} - \Omega_{\Gamma h})} \bigg/ \left(\sum_j \frac{k^2_{\Gamma j}}{(\omega_{\Gamma j} - \Omega_{\Gamma h})^2} \right)^{\frac{1}{2}} \tag{4.31}$$

one achieves the simplification in (4.20) that $d_{\Gamma hh'} = 0$ for all $h \neq 1$.

A further consequence of the choice (4.31) is that the phonon-like frequencies in (4.23) can be obtained from an algebraic equation involving only the pure lattice frequencies $\omega_{\Gamma h}$ and the coupling constants $k_{\Gamma h}$. In effect $\Omega_{\Gamma k}$ are solutions of

$$\sum_h k^2_{\Gamma h}/(\omega_{\Gamma h} - \Omega) = 0 \tag{4.32}$$

It follows from the algebraic properties of the above sum that almost all eigenvalues $\Omega_{\Gamma k}$ lie between a pair of the original frequencies $\omega_{\Gamma h}$, so that the bands of the phonons and of the phonon-like modes coincide.

In a more general use of the variational principle [51a)] one postulates a trial ground state wave-function of the form

$$\Psi = \Psi_{QM}(\tilde{q}_{\Gamma\gamma 1}) \prod_{\Gamma'\gamma'} u_0(\tilde{q}_{\Gamma'\gamma'h}) \tag{4.33}$$

where $u_0(q) \propto \exp(-\cdot\frac{1}{2} q^2)$ is a normalised harmonic oscillator wave-function. One then varies the transformation matrix in (4.16) in such a way as to make the expectation value of the Hamiltonian in (4.17) a minimum. This takes the form after simplification

$$\langle \Psi | H | \Psi \rangle = \frac{1}{2} \sum_{\Gamma\gamma} \left(\sum_h \hbar \omega_{\Gamma h} - \hbar \Omega_{\Gamma} \right) + E(\Omega_{\Gamma}, K_{\Gamma}) .$$

E is the ground state energy of H_{QM}. The variational solution for the first row of the transformation matrix can be put in the following form [51b)]:

$$A^{\Gamma}_{1h} = \frac{\omega_{\Gamma h} k_{\Gamma h}}{\Omega_{\Gamma}} \frac{\partial E}{\partial K_{\Gamma}} \left[2 (\Omega_{\Gamma} - \omega_{\Gamma h}) \left(\frac{\partial E}{\partial \Omega_{\Gamma}} - \frac{\hbar D_{\Gamma}}{2} \right) \right.$$
$$\left. + \left(\frac{2 \omega_{\Gamma}}{\Omega_{\Gamma}} - 1 \right) K_{\Gamma} \frac{\partial E}{\partial K_{\Gamma}} \right]^{-1} \tag{4.34}$$

where D_{Γ} is the dimension of the representation Γ. In the particular case of $E \otimes \sum_i \epsilon_i$, ($\Gamma = E$, $D_{\Gamma} = 2$), one can set this result in a form similar to (4.29) upon putting

$$f(\omega_{\Gamma h}) = K_{\Gamma}^{-1} \frac{1 + \xi(K_{\Gamma})}{\omega_{\Gamma h} + \Omega_{\Gamma}(K_{\Gamma})} \tag{4.35}$$

where

$$\xi(K) = \left(E - 1 - \frac{1}{2} K \frac{\partial E}{\partial K} \right) \bigg/ \left(1 - E + K \frac{\partial E}{\partial K} \right) \tag{4.36}$$

To facilitate comparison between this result and the simpler one in (4.30) for $E \otimes \Sigma_i \, \epsilon_i$, we quote the limiting forms of the ground state energy of H_{QM} for strong and weak couplings:

$$E(\kappa) = \hbar\Omega[\tfrac{1}{2}(1 - K^2) - \tfrac{1}{2}K^{-2} + O(K^{-4})] \qquad K \gg 1$$

$$\sim \hbar\Omega(1 - K^2 + \tfrac{1}{2}K^4) \qquad K \ll 1$$

In these limits $\xi(K)$ as given in (4.33) tends to zero and (4.34) coincides with (4.29). In between the two transformations differ, especially in that (4.27) diverges in the long wave-length limit, as $k_{rh} \propto \omega_{rh}^{-\frac{1}{2}}$ for small ω (this asymptotic behaviour will be further discussed in Sect. 4.4.3), whereas (4.34) tends to a finite value.

The quantity ξ leads us to the definition of l_c, a characteristic number of lattice cells, through

$$l_c^{-1} = \xi(K)\left(\frac{\Omega_a}{\mathcal{V}_L}\right) \tag{4.37}$$

$\mathcal{V}_L$ being the longitudinal velocity of sound in the crystal and a the lattice spacing. l_c will be found useful in establishing a distance dependence of the resonance mode, to which we turn now.

4.4.3 The Range of Resonance Modes

In a molecular model such as in Sect. 3.1 the modes are made up of the movements of the impurity and its nearest neighbours. In the variationally constructed resonance modes the moving atoms extend well beyond the nearest neighbours. Their range is determined by the wave-number dependent strengths of the vibronic coupling and by the phonon dispersion curve. Formally, one obtains the spatial dependence of a mode by inverting (2.2) and evaluating the displacements $u_\alpha(l\kappa)$ at the positions $l\kappa$, under restriction of the displacement pattern to the mode.

The set of 1 transformations that gives $u_\alpha(l\kappa)$ in terms of a resonance mode was given by Halperin and Englman [65c]. For the tetragonal distortion (the θ-component) of an E-mode the l-dependence of $u_\alpha(l\kappa)$ is given, for large l and within a constant factor of proportionality, by the integral

$$\int_0^\infty \frac{ds}{s + l_c^{-1}} \left[\frac{\sin\,[s(l - 1)]}{l - 1} - \frac{\sin\,[s(l + 1)]}{l + 1}\right] \, .$$

Evaluating this in limiting cases one finds

$$u_\alpha(l k) \propto l^{-2} \qquad \text{for} \qquad 1 \ll l \ll l_c \tag{4.38}$$

$$\propto l^{-3} \qquad \text{for} \qquad l_c \ll l \tag{4.39}$$

In the model of O'Brien leading to (4.29) $l_c = \infty$, since ξ which enters in (4.37) is zero [(Cf. (4.30) and (4.35))], so that from (4.38) the mode displacements decay according to the inverse square law. For the more general variational model the same power law holds in each of the limiting cases $K \gg L$ and $K \ll 1$, since in these limits $\xi \to 0$ as argued earlier. There will eventually be a very distant region $l \gg l_c = l_c(K)$ where the cubic power law of (4.39) takes over, but in this region the mode amplitudes have already substantially decreased. On the other hand, for moderate values of $K(\sim 1)$ l_c turns out to be moderately small (on the order of 5) and the spatial decay of the resonance mode is in a substantial region as l^{-3}.

It must be carefully noted that the preceding discussion has referred to the distance dependence of the resonance-mode amplitudes alone. This mode arose from the quasi-molecular Hamiltonian H_{QM} with exclusion of the perturbative residual coupling H'. Corrections due to this coupling will have only a minor effect, proportional to the mean square deviation $\langle \Delta \omega^2 \rangle$, in the region not very far from the impurity. (Characteristically for $l \lesssim 10$). However, in more distant regions the correction due to H' can be shown to become increasingly more important. This causes the distortion to fall off as l^{-2}, quite generally. Such dependence was indeed predicted early on [143] and argued to follow from general energy considerations (Stoneham 1975, Sect. 8.3.3). Here we have shown that different asymptotic regions also exist around the impurity in which the ultimate, far asymptotic behaviour of the distortion is not yet reached. This is expected to be of practical importance since, e.g. an ENDOR experiment with even the most optimistic result is unlikely to gauge the far asymptotic region.

4.4.4 The Excited ^2E_g State of Fe^{2+} in MgO

Probably the most stringent verification so far of the cluster mode description has come from the low temperature near-infrared absorption data in this system [73]. The analysis of the zero phonon line under uniaxial stress and of the associated broad absorption band led to a consistent set of parameters for a single coupled mode (K_E, Ω_E and parameters of anharmonic coupling). The derived mode frequency Ω_E agreed with the independently made estimate [115] based on the temperature dependence of the peak separation in the absorption band. At the same time the estimated coupling strength K_E was smaller than that calculated from a molecular FeO_6 cluster. Such behaviour would indeed be expected from the broadly extended cluster mode which forms the subject of this section.

5.1 Time Development of Vibrational Excitations

The following sequence of processes is common in the optical spectroscopy of impurities.

The impurity is excited by incident light from its ground state to electronically and vibrationally states. If the exciting light is pulsed so that it covers a broad spectral range, the set of states excited lie vertically above the ground in a configuration dia-

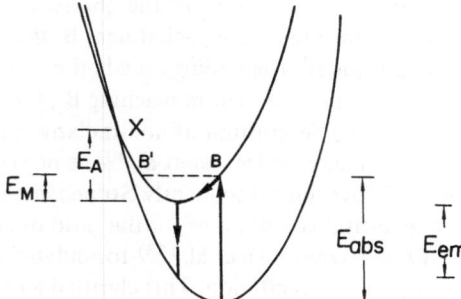

Fig. 10. Potential curves versus interaction coordinate. The evolution of the excitation during an absorption-emission cycle is shown by arrowed lines. The energy differences and lettered points are referred to in the text

gram picture. This is shown by the upward pointing line in the adjacent figure (Fig. 10). For exciting light that has a narrow frequency range a sharp vibrational level may be excited. Either of these situations forms a starting point for the time development of the excitation.

Subsequent to the excitation, the energy of the excitation will be either passed over to the surrounding lattice or emitted as light (not excluding though the possibility that partly this and partly that will happen to the excitation energy). In the former case one needs to study the histories of both the loss of vibrational energy on the impurity centre (consisting, for the purposes of this section, of the vibrating cluster or quasi-molecule of Sect. 4.2 and Sect. 4.4.2) and the diffusing-away of the energy to distant regions of the lattice. As we shall see, the experimental techniques that trace the two time-developments are essentially different.

We should emphasize that in considering the fate of the excitation on the impurity centre we are here concerned only with the vibrational component in the excitation energy (sometimes also called "the energy excess"). The purely electronic part is not of interest here and we shall only recall that the electronic de-excitation process on an impurity follows usually an energy gap law[50c,87,93]

5.2 Broad Band Excitations

We start by tracing the time-development of the vibrational excitation following a broad-band optical or UV absorption. The initially excited state will be a wave-packet residing essentially at the point B in Fig. 10 (the Franck-Condon principle).

The time-development of this excitation can be nicely understood on the basis of a wide set of data in a series of F-centres and a classical model for the motion of the wave-packet on the potential curves of Fig. 10. With numerous alkali halides and several oxides the emperical rule was found[13] that radiative emission will take place provided that the initial position of the wave packet B lies below the crossing point (denoted by X in Fig. 10) of the potential energy curves. This criterion can be expressed in terms of the quantities E_A (the activation energy) and E_M (half the Stokes shift, the relaxation energy) indicated on the figure, as $E_M < E_A$, or in terms of the experimentally available parameters E_{abs} and E_{em} (the mean energies of absorption and emission), as $2 E_{em} > E_{abs}$.

149

The physical meaning of the observed regularities in the F-centre, is that upon excitation into the wave-packet near B, this swings over to the turning point on the left hand side B′. Depending on whether the wave-packet will or will not pass through the crossing point X before reaching B′, the system will or will not decay non-radiatively. Such a description of non-radiative processes, based on simplified rules proposed originally by Dexter et al.[40], is not complete from a quantum mechanical point of view and subsequently Stoneham and Bartram[145] refined their model and gave quantitative estimates for the ratio of radiative and non-radiative decays.

Earlier Toyozawa et al.[153] formulated a dynamic model for the behaviour of the vibrational excitation. This clarified and improved upon the classical description of a non-diffusing wave-packet. Such description appears justifiable for F-centres whose optical bands are broad and diffuse not exhibiting any fine structure that might be indicative of discrete, quantised vibrational levels. (The relatively large spatial extent of the F-centre electronic wave functions also favours the involvement of long wavelength acoustic vibrations. For these a classical description can be appropriate.)

As already noted the wave packet picture envisages an oscillatory motion on the upper potential. Since the coordinate q of the resonance mode (the interaction mode, in Toyozawa's terminology) is not a proper stationary mode, but rather a superposi-

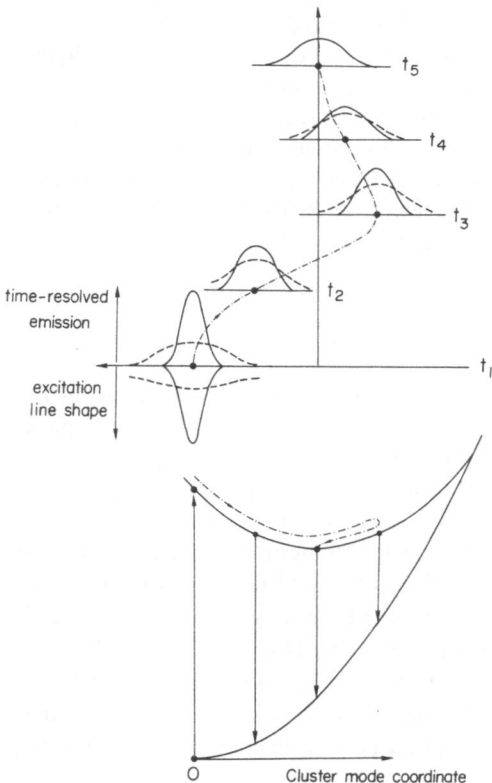

Fig. 11. The motion of the excitation wave-packet (after Toyozawa et al.[153]). The lower part of the figure exhibits the excitation pathway in the space of the cluster mode coordinate, including the overshooting (once) of the minimum. In the upper part, the time-development of the energy of excitation (dash-dotted curve) is tracted, with energy increasing along the horizontal arrow and time along the vertical one. The solid and broken curves represent the cases of nearly monochromatic and nearly white excitations

tion of eigenmodes, the interference between these will tend to damp out the oscillation. Such damped oscillation is depicted by broken lines in Fig. 11. It is then proposed that the damping in the F-centres is characteristically weak or that the overshooting of the minimum, towards the crossing point and beyond, is complete. If so, in this dynamic description the criterion for emission is identical to that in the classical picture.

A formal treatment can also be provided. Returning to the situation, an $A_{1g} \rightarrow T_{1u}$ transition, described in Sect. 4.2.2 which is indeed analogous to some bands in F-centres, and restricting ourselves to one component (say, T_{1uz}) of the excited state and to the symmetric $Q_{A_1}(\equiv Q_A)$ only, the *instantaneous* energy difference (i.e. that for a fixed value of the coordinate) can be written by following (4.14).

$$\Delta H = H_{T_{1uz}} - H_{A_{1g}} = \Delta E - V_A M_A^{-\frac{1}{2}} Q_A(t) . \tag{5.1}$$

Here the displacement coordinate $Q_A(t)$ is given a time dependence since we are adopting a classical description. As explained at length in Sect. 4.4.2, in a lattice the role of the molecular coordinate Q_A is taken by the corresponding resonance mode coordinate which is a superposition of the pure lattice normal coordinates $q_{A_i}(\equiv q_i$, $i = k^* jn)$ defined in (3.13). In terms of these and the electron-phonon coupling coefficients in (3.15) $k_{Ai}(\equiv k_i)$, (5.1) can be rewritten as

$$\Delta H = \Delta E - \sum_i \hbar \omega_i k_i q_i(t) . \tag{5.2}$$

In the excited state the mode i oscillates about the equilibrium position located at $\frac{1}{2} k_i$ and has at time t = 0 the value $-\frac{1}{2} k_i$. Therefore its time development follows

$$q_i(t) = \frac{1}{2} k_i(1 - 2 \cos \omega_i t)$$

and

$$\Delta H = (\Delta E - \frac{1}{2} \sum_i \hbar \omega_i k_i^2) + \sum_i \hbar \omega_i k_i^2 \cos \omega_i t \tag{5.3}$$

The constant term in parentheses is just the energy difference between the two curves at the equilibrium point on the upper curve. The excess energy ΔH describes oscillations about this value. If all frequencies ω_i have the same value then (5.3) exhibits true harmonic behaviour of the energy difference between two extreme values. By indication of the experimental data such behaviour appears to prevail in F-centres, although this cannot be exactly true, as we have already remarked.

In a more general manner, the frequency spread, $\langle \Delta \omega^2 \rangle$ in (4.27), is non-zero, (5.3) is not a single harmonic and the overshooting of the minimum will not have its full extent. This is shown in Fig. 11. The damping-out of the energy difference is also shown by broken lines on the upper curve. The damping will become significant after a time of the order of

$$\langle \Delta \omega^2 \rangle^{-\frac{1}{2}},$$

this being the dephasing time of the initial excitation. If the impurity mode Q_A is compared (fully or partly) of a finite (as opposed to infinitesimal) component in one stationary mode (as in e.g. a localised mode), a finite oscillatory component would remain in ΔH.

5.3 Development of Excitations in Sharp Vibrational Levels

Quantised vibrational levels in impurity centres may arise from localised or fairly long-lived resonance modes or from internal modes of a molecular impurity[2,44,63,95]. For the former situation the development was traced theoretically for several decades of the time[10] while the exchange of excitation among the vibrational levels of a molecular oscillator was the subject of some now classic papers in statistical mechanics[17,107,169]. For a molecular impurity in which a vibrational level is initially selectively excited, the occupation probability of this level will decrease in time due to its feeding of other levels. If the excitation remains on the molecule, other levels of the molecule will gain in occupation, until ultimately a steady state situation (usually, of thermal equilibrium) will prevail amidst the molecular degrees of freedom. The story becomes more complicated if there is also an exchange of energy with the surrounding lattice[55]. Then there are likely to be two characteristic relaxation times, a shorter one leading to intra-molecular equilibrium and a longer time for thermalisation with the lattice.

A fertile area of experimentation is the selective excitation of vibrational levels of small to medium sized molecules in lattices of rare gas atoms[22]. The molecules

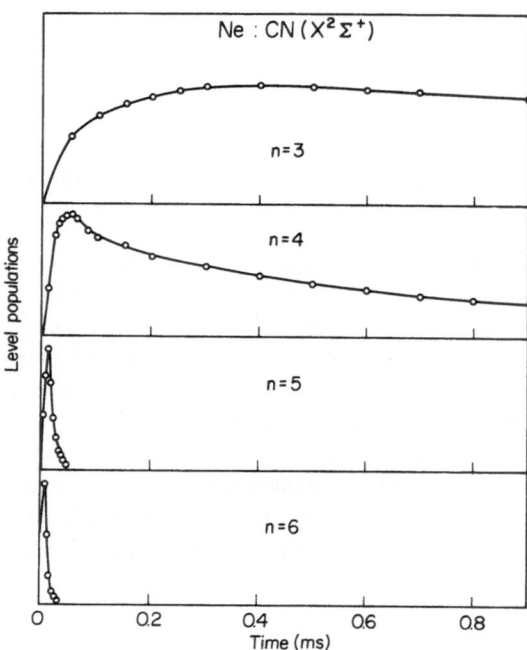

Fig. 12. Population of excited vibrational levels in the ground electronic state ($X\,^2\Sigma^+$) of Ne:CN as function of time. Data points (circles) due to Bondybey and Nitzan[24] were obtained by laser pulse excitation in the excited state A $^2\Pi$ levels and probing of the level populations in the ground state by a second laser

range from the diatomics (OH, HCl, N_2, NO, etc.) to larger kinds (e.g. NH_3, CH_3F). The lifetimes of the vibrational excitations are generally between 10^{-2} and 10^{-8} seconds. The observed history of excitations in a particular case is shown in Fig. 12. This is essentially what one expects from a set of interacting discrete vibrational levels.

The de-excitation of a vibrational level may be accompanied by an electronic excitation or else the newly excited level may be a lower lying level of the same electronic state. Some interesting situations were found for C_2, C_2^- and CN [21] in rare gas matrices, where the excitation zig-zags between vibrational levels, now of one then of another vibrational origin, favouring as a rule the nearest lower level, irrespective of the type of vibration excited.

In a broad family of cases the excited vibration loses its energy to the lattice in two steps. In the first step, which is frequently rate determining, it excites localised vibrations, in particular a rotational type of motion of the guest molecule or a translational motion of the same in the lattice cage [19]. Secondly the localised excitation diffuses away to the farther reaches of the lattice.

Little work has been accomplished in this topic, either by experiments or theoretically. From the point of the point of the present Report, this is clearly a situation that wants to be remedied, since the diffusion of the localised vibrational excitation is a key property in the understanding and utilisation of the impurity-modes.

5.4 Phonon-Pulse Spectroscopy

The trajectory of the vibrational excitation into the crystal can be detected by a microphone or a bolometer which monitors the pressure fluctuation associated with

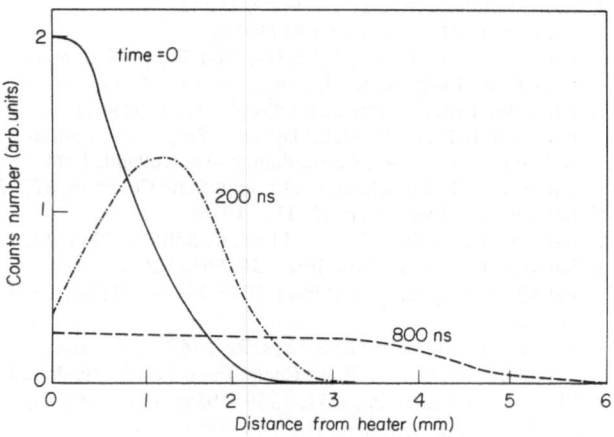

Fig. 13. Spatial distribution of the $29\,cm^{-1}$ vibrational excitation on Cr^{3+} in ruby at three times after heat injection (after Renk [130]). The heat injector, having a time resolution of about 10 ns, perturbs the population within the ^2E-levels. The position of the phonon detector is then shifted, with a spatial resolution of about 1 mm, relative the light excited region of the crystal

the vibration or heat-pulse. This is a currently developing and greatly sensitive method, based upon an idea of A.G. Bell (the inventor of the telephone). In one design, that was used by Robin and Kuebler[132] to monitor the chromium decay in ruby, a response time of 10^{-7} sec and a temperature sensitivity of 10^{-5} degrees was achieved by means of a thin superconducting lead film in close contact with the sample. The technique has obvious potentialities for our understanding of vibrational excitations near impurities. Reviews have been written by Rosencwaig[133] and Colles et al.[34].

An optical detection technique was developed by Renk[130] to observe the distribution of vibrational excitations in a ruby sample during 800 nanoseconds after injecting a heat pulse. A spatial resolution of about 1 mm and a 10^{-9} sec time response were achieved. The result is shown in Fig. 13. The orientational distribution of the pulse was also studied[83]. From later work[120] it has become clear that in Fig. 13 the 29 cm^{-1} phonon, emanating from the excited ^{2}E-state in ruby, propagates ballistically rather than as a post-relaxation heat wave.

The experimental difficulties in the generation of monochromatic phonons in the wave-number range above 10 cm^{-1} seem to have been overcome[42,102].

6 References

1. Abou-Ghantous, M. et al.: J. Phys., C: Solid St. Phys. 7, 2707 (1974)
2. Abouaf-Marguin, L., Gauthier-Roy, B., Legay, F.: Chem. Phys. 23, 443 (1977)
3. Abragam, A.A., Bleaney, B.: Electron Paramagnetic Resonance of Transition Metal Ions, Oxford, University Press 1967
4. Alexander, R.W., Hughes, A.E., Sievers, A.J.: Phys. Rev. B 1, 1563 (1970)
5. Allen, J.W.: J. Phys., C: Solid St. Phys. 1, 1130 (1968)
6. Ashkin, M.: J. Phys. (Paris) 26, 709 (1965)
7. Bacci, M.: Phys. Stat. Solidi (b) 82, 169 (1977)
8. Bardeen, J.: Phys. Rev. 52, 688 (1937)
9. Barker, A.S. Jr., Sievers, A.J.: Rev. Mod. Phys. 47, Supplement S1-S179 (1975)
10. Barnett, B., Englman, R.: J. Luminescence 3, 55 (1970)
11. Barth, W., Fritz, B.: Phys. Stat. Solidi 11, 515 (1967)
12. Barron, T.H.K., Klein, M.L.: Dynamic Properties of Solids Vol. I, (Horton G., Maradudin, A.A. ed.) p. 391–449, Amsterdam, North Holland, 1974
13. Bartram, R.H., Stoneham, A.M.: Solid State Commun. 27, 1593 (1975)
14. Bates, C.A.: Physics Rep. 35, 187 (1978)
15. Bates, C.A., Wardlaw, R.S.: J. Phys., C: Solid St. Phys. 12, 2133 (1979)
16. Bauman, F.C. et al.: Phys. Rev. 159, 691 (1967)
17. Belavin, A.A. et al.: Soviet Phys. JETP 29, 145 (1968), Zh. Eksperim. Teor. Fiz. 56, 264 (1968)
18. Benedek, G., Nardelli, G.F.: Phys. Rev. 167, 837 (1968)
19. Berkowitz, M., Gerber, R.B.: Chem. Phys. Letters 49, 260 (1977)
20. Blasse, G.: J. Chem. Phys. 51, 3529 (1969); J. Solid St. Chem. 9, 147 (1974)
21. Bondybey, V.E.: J. Chem. Phys. 66, 995 (1977)
22. Bondybey, V.E., Brus, L.E.: Adv. Chem. Phys. (to appear)
23. Bondybey, V.E., English, J.M.: J. Chem. Phys. 68, 4641 (1978)
24. Bondybey, V.E., Nitzan, A.: Phys. Rev. Lett. 38, 889 (1977)
25. Born, M., von Karman, Th.: Phys. Zeitschr. 13, 297 (1912)
26. Brafman, O. et al.: Phys. Rev. Lett. 19, 1120 (1967)

27. Brout, R., Visscher, W.: Phys. Rev. Lett. *9*, 54 (1962)
28. Caldwell, R.F., Klein, M.V.: Phys. Rev. *158*, 851 (1967)
29. Cochran, W.: a) Phil. Mag. *4*, 1082 (1959); b) Proc. Phys. Soc. (London) A*253*, 260 (1959)
30. Cowley, R.A. et al.: Phys. Rev. *131*, 1030 (1963)
31. Chase, L.L., Hao, C.H.: Phys. Rev. B *12*, 5990 (1975)
32. Chase, L.L., Kuhner, D., Bron, W.E.: Phys. Rev. B 7, 3892 (1973)
33. Choquard, P.: The Anharmonic Crystal, New York, Benjamin 1967
34. Colles, M.J., Geddes, N.R., Mehdizadeh, G.: Contemporary Physics, *20*, 11 (1979)
35. Dang, L.S. et al.: Phys. Rev. B *18*, 2989 (1978)
36. Dawber, P.G., Elliott, R.J.: a) Proc. Phys. Soc. (London) *81*, 453 (1963); b) ibid. *273*, 222 (1963)
37. Dean, P.J., Guthbert, J.D., Lynch, R.T., Phys. Rev. *179*, 754 (1969)
38. Debye, P.: Ann. Phys. (Leipzig) *39*, 789 (1912)
39. De Jong, C., van der Elsken, J.: in *Phonons* (Nusimovici, M.A., ed.) p. 391, Paris, Flammarion, 1971
40. Dexter, D.L., Klick, C.C., Russell, G.A.: Phys. Rev. *100*, 603 (1955)
41. Di Bartolo, B., Peccei, R.: Phys. Rev. *137A*, 1770 (1965)
42. Dijkhuis, J.I., van der Pol, A., de Wijn, M.W.: Phys. Rev. Lett. *37*, 1554 (1976)
43. Dolling, G. et al.: Phys. Rev. *147*, 577 (1966)
44. Dubost, H., Charneau, R.: Chem. Phys. *12*, 407 (1976)
45. Duke, C.B., Mahan, G.D.: Phys. Rev. *139A*, 1965 (1965)
46. Elliott, R.J.: in *Phonons* (Stevenson, R.W.H., ed.) Chap. 14, New York, Plenum, 1966
47. Elliott, R.J., Gibson, A.F.: An Introduction to Solid State Physics and its Applications, London, MacMillan, 1974
48. Elliott, R.J., Krumhansl, I.A., Leath, P.L.: Rev. Mod. Phys. 46, 465 (1974)
49. Elliott, R.J., Parkinson, J.B.: J. Phys. Soc (London) *92*, 1024 (1972)
50. Englman, R.: a) Mol. Phys. *3*, 23 (1961); b) The Jahn-Teller Effect in Molecules and Crystals, London, Wiley, 1972; c) Non-radiative Decay of Ions and Molecules in Solids, Amsterdam, North Holland, 1979
51. Englman, R., Halperin, B.: a) J. Phys. C: Solid St. Phys. *6*, L 219 (1973); b) Ann. Phys. (Paris) *3*, 453 (1978)
52. Enns, R.H., Haering, R.R. (Ed.): Modern Solid State: Vol. 2 Phonons and Their Interactions, New York, Gordon and Breach, 1969
53. Faulkner, R.A., Dean, P.J.: J. Luminescence *1*, *2*, 552 (1970)
54. Feldman, D.W., Ashkin, M., Parker, J.H. Jr.: Phys. Rev. Lett. 17, 1209 (1966)
55. Fischer, S.F.: ibid. *17*, 25 (1972)
56. Fitchen, D.B.: in *Physics of Color Centers* (Fowler, W.B., ed.) p. 293–350, New York, Academic Press, 1968
57. Fletcher, J.R.: J. Phys. C: Solid St. Phys. *5*, 852 (1972)
58. Fletcher, J.R., Knowles, A., Moore, W.S. (to appear)
59. Fowler, W.B.: Physics of Color Centers, New York, Academic Press, 1968
60. Gauthier, N., Walker, M.B.: Phys. Rev. Lett. *31*, 1211 (1973); Canad. J. Phys. *54*, 9 (1976)
61. Gebhardt, W., Kühnert, H.: Phys. Letters *11*, 15 (1964); Phys. Stat. Solidi *14*, 157 (1966)
62. Geschwind, V.W., Remeika, J.P.: J. Appl. Phys. *33*, 370 (1972)
63. Goodman, J., Brus, L.E.: J. Chem. Phys. *95*, 1853 (1978)
64. György, E.M. et al.: Phys. Rev. Lett. *15*, 19 (1965)
65. Halperin, B., Englman, R.: a) Phys. Rev. Lett. *31*, 1052 (1973); b) Phys. Rev. *B 9*, 2264 (1974); c) ibid. *B 12*, 366 (1975); d) J. Phys. C: Solid State Phys. *8*, 3975 (1975); e) J. Luminescence *19*, 229 (1979)
66. Ham, F.S.: Electron Paramagnetic Resonance (Geschwind, S., ed.) p. 1, New York, Plenum Press, 1972
67. Ham, F.S., Schwarz, W.M., O'Brien, M.C.M.: Phys. Rev. *185*, 548 (1969)
68. Hardy, J.R., Karo, A.M.: The Lattice Dynamics and Statics of Alkali Halide Crystals, New York, Plenum Press, 1979
69. Hattori, T. et al.: Solid State Commun. *12*, 545 (1973)

70. Hayes, W., Wiltshire, M.C.K., Dean, P.J.: J. Phys. C: Solid St. Phys. *3*, 1762 (1970)
71. Henderson, B.: *Defects in Crystalline Solids*, London, Edward Arnold, 1972
72. Henry, C.H., Slichter, C.P., in: *Physics of Color Centers* (Fowler, W.B., ed), New York, Academic Press, 1968
73. Hjortsberg, A. et al.: Phys. Rev. Lett. *39*, 1233 (1977)
74. Horner, M.: Dynamical Properties of Solids (Horton, G.K., Maradudin, A.A., eds.) Vol. I, p. 452–498, Amsterdam, North Holland, 1974
75. Huang, K. Rhys, A.: Proc. Roy. Soc (London) *A 204*, 406 (1950)
76. Hughes, A.E.: Proc. Phys. Soc. (London) *87*, 535 (1966), *88*, 449 (1966)
77. Hughes, A.E. et al.: AERE Rep. R 5606 (1967)
78. Imry, Y., Wortis, M.: Phys. Rev. *B19*, 3580 (1979)
79. Irwin, J.C., Clayman, B.P., Mead, D.G.: Phys. Rev. *B 19*, 2089 (1979)
80. Jones, W., March, N.H.: *Theoretical Solid State Physics*, Vols. 1 and 2, London, Wiley, 1973
81. Kaiser, W.: J. Phys. Chem. Solids *23*, 255 (1962)
82. Kagan, Yu. M., Iosilevskii, Ya. A.: Soviet Phys. JETP 15, 182 (1962); Zh. Eksperim. Teor. Fiz. *42*, 259 (1962)
83. Kaplyanskii, A.A.: Spectroscopie des Elements de Transition et des Elements Lourds dans les Solides, p. 137–141, Paris, Editions CNRS, 1977
84. Karlov, N.V., Margerie, J., Merle d'Aubigne, Y.: J. Phys. (Paris) *24*, 717 (1963)
85. Karo, A.M., Hardy, J.R.: Phys. Rev. *129*, 2024 (1963); ibid. *141A*, 696 (1966)
86. Keeler, G.J., Batchelder, D.N.: J. Phys. C: Solid St. Phys. *5*, 3264 (1972)
87. Kiel, A.. *Quantum Electronics* (Grivet, P., Bloembergen, N., eds.) p. 765, Paris, Dunod, 1964
88. Kittel, C.: Quantum Theory of Solids, New York, Wiley, 1963
89. Klein, M.V.: Physics of Color Centers (Fowler, W.B., ed.) p. 429–535, New York, Academic Press, 1968
90. Koster, G.F., Slater, J.C.: a) Phys. Rev. *95*, 1167 (1954); b) Phys. Rev. *96*, 1208 (1954)
91. Klemens, P.G.: Solid St. Phys. *5*, 495 (1958)
92. Kubo, R., Nagamiya, T.: *Solid State Physics*, New York, McGraw-Hill, 1969
93. Kubo, R., Toyozawa, Y.: Progr. Theor. Phys. *13*, 161 (1955)
94. Kühner, D.M., Lauer, H.V., Bron, W.E.: Phys. Rev. *B5*, 4112 (1972)
95. Legay, F.: *Chemical and Biochemical Applications of Lasers*, (Moore, C.B., ed.) Vol. 2, p. 43, New York, Academic Press, 1977
96. Lombardo, G., Pohl, R.D.: Phys. Rev. Lett. *15*, 291 (1965)
97. Maradudin, A.A., Caldwell-Horsfall, R.A.: J. Phys. (Paris) *26*, 717 (1965)
98. Maradudin, A.A., Fein, A.F.: Phys. Rev. *128*, 2589 (1962)
99. Maradudin, A.A. et al.: Theory of Lattice Dynamics in the Harmonic Approximation, 2nd Ed., New York, Academic Press, 1971
100. McCumber, D.E., Sturge, M.D.: J. Appl. Phys. *34*, 1682 (1963)
101. McPherson, R.W., Timusk, T.: Can J. Phys. *48*, 2176 (1970)
102. Meltzer, R.S., Rives, J.E.: Phys. Rev. Lett. *38*, 421 (1977)
103. Messmer, R.P., Watkins, G.D.: Phys. Rev. Lett. *25*, 656 (1970); Phys. Rev. *B7*, 2568 (1973)
104. Mirlin, D.N., Reshina, I.I.: Soviet Phys. Solid State *6*, 2454 (1964); Fiz. Tverdogo Tela *6*, 3078 (1964)
105. Montgomery, G.P. et al.: Phys. Rev. *B6*, 4047 (1972)
106. Montroll, E.W., Potts, R.W.: Phys. Rev. *100*, 525 (1955)
107. Montroll, E.W., Shuler, K.E.: J. Chem. Phys. *26*, 454 (1957)
108. Muramatsu, S.: Solid State Commun. *21*, 125 (1977)
109. Nakamoto, K.: Infrared Spectra of Inorganic and Coordination Compounds, New York, Wiley, 1963
110. Newman, R.C.: Physics, *18*, 545 (1969)
111. Newman, R.C., Smith, R.S.: Phys. Lett. *A24*, 671 (1967)
112. Newman, R.C. et al.: Phys. Lett. *33A*, 113 (1970)
113. Nguyen Xuan Xinh, Maradudin, A.A., Coldwell-Horsfall, R.A.: J. Phys. (Paris) *26*, 717 (1965)
114. Nüsslein, V., Schröder, V.: Phys. Stat. Solidi *21*, 309 (1967)
115. Nygren, B.: Ph. D. Thesis, Chalmers University 1975 (quoted by Hjoltsberg et al., 1977)

116. O'Brien, M.C.M.: J. Phys. C: Solid State Phys. *5*, 2045 (1972)
117. Okazaki, M. et al.: J. Phys. Soc. Japan, *22*, 1349 (1967)
118. Orbach, R.: Optical Properties of Ions in Solids, p. 355 (Di Bartolo, B., ed.) New York, Plenum Press 1975; J. Luminescence *18/19*, 634 (1979)
119. Orbach, R., Stapleton, H.J.: Electron Paramagnetic Resonance, p. 121 (Geschwind, S., ed.) New York, Plenum Press, 1972
120. Pauli, G., Renk, K.F.: Phys. Lett. *67A*, 410 (1978)
121. Paulusz, A.G.: J. Luminescence *17*, 375 (1978)
122. Peascoe, J.G., Fenner, W.R., Klein, M.V.: J. Chem. Phys. *60*, 4208 (1974)
123. Pekar, S.I.: Zh. Eksperim. Teor. Fiz. *20*, 519 (1950)
124. Perlin, Yu. E.: Soviet Phys. Usp. *6*, 542 (1964); (Uspekhi Fiz. Nauk. *80*, 553 (1963))
125. Prevot, B., Carabatos, C., Schwab, C.: Solid State Commun. *28*, 964 (1973)
126. Radlinski, A.B.: J. Luminescence *18/19*, 147 (1979)
127. Rebane, K.: *Impurity Spectra of Solids*, New York, Plenum Press 1976
128. Rebane, K., Saari, P.: J. Luminescence *12–13*, 23 (1976)
129. Reisfeld, R.: *Structure and Bonding 13*, 53 (1973)
130. Renk, K.F.: Festkörperprobleme XII (Madelung, O., ed.) p. 107–132, Oxford, Pergamon Press, 1972
131. Rivallin, J., Salce, B., de Goer, A.M.: Solid State Commun. *19*, 9 (1976)
132. Robin, M.B., Kuebler, N.A.: J. Chem. Phys. *66*, 169 (1977)
133. Rosencwaig, A.: Electronics and Electron Physics *46*, 207 (1978)
134. Salce, B., de Goer, A.M.: J. Phys. C: Solid St. Phys. *12*, 2081 (1979)
135. Schäfer, G.: J. Phys. Chem. Solids *12*, 233 (1960)
136. Seitz, F.: J. Chem. Phys. *6*, 150 (1938)
137. Shen, L.N., Estle, T.L.: a) J. Phys. C: Solid St. Phys. *12*, 2103 (1979); b) ibid. *12*, 2119–2130 (1979)
138. Sievers, A.J.: Phys. Rev. Lett. *13*, 310 (1964)
139. Simonetti, J., McClure, D.S.: Phys. Rev. *16*, 3887 (1977)
140. Spitzer, W.G.: Festkörperprobleme XI, p. 1–44, (Madelung, O., ed.) Oxford, Pergamon Press, 1971
141. Spitzer, W.G. et al.: J. Appl. Phys. *40*, 2589 (1969)
142. Steggles, P.: J. Phys. C: Solid St. Phys. *10*, 2817 (1977)
143. Stevens, K.W.H.: J. Phys. C: Solid St. Phys. *2*, 1934 (1969)
144. Stoneham, A.M.: *Theory of Defects in Solids*, Oxford, Clarendon, 1975
145. Stoneham, A.M., Bartram, R.M.: AERE Repr. TP 740 (1978)
146. Sturge, M.D.: Phys. Rev. *131*, 1456 (1963), *140A*, 880 (1965); Solid State Phys. *20*, 91 (1967)
147. Sturge, M.D. et al.: Phys. Rev. *155*, 218 (1967)
148. Takeno, S.: Progr. Theor. Phys. *29*, 191 (1963)
149. Taylor, D.W.: Dynamical Properties of Solids (Horton, G.K., Maradudin, A.A., eds.), Vol. 11, p. 285–384, Amsterdam, North-Holland, 1975
150. Thomas, H.A.: Private communication 1978
151. Toyozawa, Y.: Dynamical Processes in Solid Optics (Kubo, R., Kamimura, H., eds.), p. 90–115, Tokyo, Syokabo, 1967
152. Toyozawa, Y. et al.: J. Phys. Soc. Japan *22*, 1337 (1967)
153. Toyozawa, Y., Kotani, A., Sumi, A.: J. Phys. Soc. Japan, *42*, 1405 (1977)
154. Trifonov, E.D., Peuker, K.: J. Phys. (Paris) *26*, 738 (1965)
155. Van Kranendonk, J.: Modern Solid State Physics, Vol. 2 (Enns, R.H., Haering, R.R., eds.), p. 107–204, New York, Gordon and Breach, 1969
156. Van Vleck, J.H.: a) Phys. Rev. *57*, 426 (1940); b) ibid. *57*, 1052 (1940)
157. Vardeny, Z., Gilat, G., Moses, D.: Phys. Rev. B. *18*, 1487 (1978)
158. Walker, C.T., Pohl, R.O.: Phys. Rev. *131*, 1433 (1963)
159. Ward, R.W., Clayman, B.P.: Phys. Rev. *B9*, 4455 (1974)
160. Ward, R.W., Clayman, B.P., Timusk, T.: Canad. J. Phys. *53*, 424 (1975)
161. Ward, R.W., Timusk, T.: Phys. Rev. *B5*, 2351 (1972)

162. Watkins, G.F., Messmer, R.P.: Phys. Rev. Lett. *32*, 1244 (1974)
163. Watts, R.K.: *Optical Properties of Solids* (Di Bartolo, B., ed.) New York, Plenum Press, 1975
164. Weber, R., Nette, P.: Phys. Rev. Lett. *20*, 493 (1966)
165. Weber, W.: Phys. Rev. *B8*, 5082 (1973)
166. Williams, F.E.: J. Chem. Phys. *19*, 457 (1951)
167. Worlock, J.M., Porto, S.P.S.: Phys. Rev. Lett. *15*, 697 (1965)
168. Yamada, N., Shionoya, Kushida, T.: J. Phys. Soc. Japan *32*, 1577 (1972)
169. Zeldovich, B.Ya., Perelemov, A.M., Popov, V.S.: Sov. Phys. JETP, *28*, 308 (1969); (Zh. Eksperim. Teor. Fiz. *55*, 589 (1968))
170. Zubarev, D.N.: Sov. Phys. Usp. *6*, 320 (1960)

Actinide-Specific Sequestering Agents and Decontamination Applications

Kenneth N. Raymond and William L. Smith*

Department of Chemistry and Materials and Molecular Research Division, Lawrence Berkeley Laboratory, University of California, Berkeley, CA 94720, U.S.A.

Table of Contents

---

* Please address all correspondence to Kenneth N. Raymond.

K. N. Raymond and W. L. Smith

I. Introduction

With the commercial development of nuclear reactors, the actinides have become important industrial elements. A major concern of the nuclear industry is the biological hazard associated with nuclear fuels and their wastes[1,2]. As seen in Table 1, the *acute chemical* toxicity of tetravalent actinides, as exemplified by Th(IV), is similar to Cr(III) or Al(III). However, the acute toxicity of ^{239}Pu(IV) is similar to strychnine, which is much more toxic than any of the non-radioactive metals such as mercury. Although the more radioactive isotopes of the transuranium elements are more acutely toxic by weight than plutonium, the acute toxicities of ^{239}Pu, ^{241}Am, and ^{244}Cm are nearly identical in radiation dose, $\sim 100~\mu$Ci/kg in rodents[6]. Thus, the extreme acute toxicity of ^{239}Pu is attributed to its high specific activity of alpha emission[6-8].

Unlike organic poisons, biological systems are unable to detoxify metal ions by metabolic degradation. Instead, unwanted metal ions are excreted or immobilized[9]. Unfortunately, only a small portion of absorbed tetra- or trivalent actinide is eliminated from a mammalian body during its lifetime. The remaining actinide is distributed throughout the body but is especially found fixed in the liver and in the skeleton [6,8,10-13]. While the ability of some metals to do damage is greatly reduced by immobilization, local high concentrations of radioactivity are produced by immobilized actinides — thereby increasing the locally absorbed radiation dose and the carcinogenic potential. Thus the long-term, chronic toxicity is much greater than the immediate, acute toxicity.

Table 1. A comparison of the acute toxicity of some chemical substances in mice [a,b]

Substance	$LD_{50/30}$, mmole/kg	Relative toxicity
NaCl	44.52	1
$CaCl_2$	2.50	18
$ZrOCl_2$	0.96	46
$CrCl_3$	0.90	49
$ThCl_4$	0.89	50
$AlCl_3$	0.80	56
$Fe_2(SO_4)_3$	0.42	106
Pb (acetate)$_2$	0.37	120
$ZnCl_2$	0.18	247
TlCl	0.10	445
$CdSO_4$	0.033	1349
UO_2Cl_2	0.021	2145
$HgCl_2$	0.020	2283
^{239}Pu(IV)Citrate	0.0047 (rat) 0.0013 (dog)	9400
Strychnine	0.0015	30000
Botulinus Toxin A	3×10^{-9} mg/kg	

a   Data for Pu from Refs. [4] and [5], organic poisons from Ref. [5], all others from Ref. [3]

b   Note that this is to be distinguished from the chronic or long-term toxicity of such substances

Primarily through the induction of bone cancer or tumors of blood forming tissue, very low doses of ^{239}Pu significantly shorten the life span of laboratory animals[5,6,8,14-17]. While mice suffered no ill effects from plutonium doses less than 1/1000 of the acutely toxic dose ($\sim$ 1 μg/kg)[17], a dose of 0.26 μg/kg given to dogs increased the incidence of bone cancer from 1/10,000 to 1/3 and decreased their lifespan 14%[17]. Lung cancer formed in all dogs that inhaled $\sim$ 1 μg/kg of plutonium oxide, but their lifespan was not significantly shortened[18]. Removal of very small amounts of actinides from the body is therefore an essential component of treatment for actinide contamination, particularly Pu(IV).

II. Biochemistry of Plutonium

While not the most toxic, plutonium is the most likely transuranium element to be encountered. In addition to the several kilograms of naturally occurring plutonium, about 5,000 kg of plutonium has been released during nuclear weapons testing, accidental destruction of nuclear devices, and nuclear fuels reprocessing[19-21]. Fortunately, the viable routes of plutonium contamination are limited to direct physical transport, since the inability of plutonium to cross physiological membranes prevents its concentration in the food chain[20-22]. The concentration of plutonium in plants is 10^{-4} to 10^{-6} of the surrounding soil[23]. Further, only 0.03% of ingested Pu(IV) citrate is absorbed by the gastrointestinal tract of mammals, while much smaller amounts of less stable chelates, simple salts, or insoluble compounds of plutonium are absorbed[10,11,13,24]. Similarly, insignificant amounts of plutonium are absorbed through intact skin during long exposures to highly acidic plutonium solutions[25]. Thus human contamination by environmental plutonium would seem to be limited to the direct ingestion or inhalation of plutonium resuspended from soil. However, there continues to be concern that naturally-occurring chelating agents might complex plutonium sufficiently strongly to change this view. Occupationally, plutonium has gained admittance to humans principally through inhalation and wounds[26].

The biological behavior of plutonium is dependent on the chemical form. Insoluble compounds of plutonium, such as oxides, fluorides, and hydroxides, largely remain in the lung or at the site of an intramuscular wound. Particles of these insoluble compounds may be slowly transported to the lymph nodes, and a small portion may react with biological ligands to form soluble complexes that are transported by the circulatory system[10,27-29]. Extremely small particles of PuO_2 when inhaled as aerosols are rapidly absorbed from the lung and enter the circulation as low molecular-weight complexes[30]. Plutonium chelates are quickly and completely absorbed from the site of entry, but metabolically inert complexes, such as Pu-DTPA, are rapidly and nearly quantitatively excreted. Complexes of metabolizable ligands, such as citrate and ascorbate, are not excreted, but give up their plutonium to plasma proteins. Other compounds of plutonium such as hydrolyzable chelates and simple salts are partially absorbed into the circulation. Much larger amounts of Pu(III) and

161

Pu(VI), which hydrolyze less readily than Pu(IV), are absorbed. The remainder hydrolyzes to form an insoluble deposit, which behaves as described above[10,11,27,31].

Particularly in the case of plutonium hydroxide, the amount of plutonium solublized from an internal deposit by biological ligands depends upon the oxidation state of the deposited plutonium. The charge to ionic-radius ratio and the tendency towards hydrolysis decreases in the order[32,33];

$$Pu^{4+} > PuO_2^{2+} > Pu^{3+} > PuO_2^+.$$

In the absence of chelating agents, hydrolysis of Pu(IV), a strong Lewis acid, occurs rapidly at low pH. Ultimately, insoluble colloids and polymers of $Pu(OH)_4$ are formed. A 4×10^{-3}M solution of Pu(IV) at pH = 1 was 40% polymerized in 30 min[34], but pseudocolloids of $Pu(OH)_4$ did not form in a 6.8×10^{-8}M solution of Pu(IV) until the pH was raised to 2.8 and polymerization was not complete below pH = 7.5[35]. The redissolution of Pu(IV) hydroxides proceeds slowly, even in the presence of chelating agents[36,37]. Because of their decreased acidity, Pu(III) and Pu(VI) hydrolyze less readily than Pu(IV). Hydrolysis of a 10^{-9}M solution of Am(III) or Cm(III) begins at pH = 4.5, and colloidal species form near pH = 7[38]. The solubility product of $Pu(OH)_3$ (2×10^{-20}) is much greater than that of $Pu(OH)_4$ (7×10^{-56})[39]. Thus, the amount and the rate of dissolution of an insoluble plutonium deposit increases in the order; Pu(III) > Pu(VI) > Pu(IV).

While the hydrolytic behavior of the oxidation state determines the amount and the rate of plutonium absorbed from the lungs or from a wound, the tissue distribution of the absorbed Pu was indistinguishable when Pu(III), Pu(IV), or Pu(VI) was administered to rats[40]. Thus, once plutonium enters the circulation its behavior is independent of its orginal oxidation state. Biologically plutonium behaves like thorium[41,42], which is stable in solution only as a tetravalent ion. While the biological behavior of Am(III) and Cm(III) is similar to plutonium, there are significant differences in the binding to endogenous ligands, biological transport, distribution, and rate of elimination[6,10,11,43]. In contrast to tri- and tetravalent actinides, the oxocations, as exemplified by the uranyl ion, are rapidly absorbed from lungs and wounds, and the majority of the absorbed uranium is rapidly excreted as an uranyl-bicarbonate complex[44-46].

In aqueous solutions each of the oxidation states of plutonium from III to VI coexist in equilibrium, and depending on the conditions and relative concentrations of the oxidation states disproportionation may occur. The redox behavior of ^{239}Pu is further complicated by its high specific activity of alpha radiation. The radiolytic decomposition of water produces oxidants (HO and HO_2 radicals, H_2O_2) and reductants (H_2O_2, H radicals), which may oxidize or reduce plutonium, depending on the relative proportions of the different valency states initially present[47]. However, Pu(III) is oxidized to Pu(IV) by water at neutral pH and Pu(III) hydroxide is rapidly oxidized by air to Pu(IV) hydroxide; while Pu(VI) is reduced to Pu(IV) by Fe(II)[33]. Further, the complexing ability of plutonium decreases in the order[48-50];

$$Pu^{4+} > Pu^{3+} \approx PuO_2^{2+} > PuO_2^+.$$

Although there is no direct measurement of the oxidation state of plutonium in biological fluids, redox potentials, complexation and hydrolysis strongly favor Pu(IV) as the dominant specie.

Plutonium which is absorbed into the circulatory system of mammals, either by injection of a metabolizable complex or by solubilization of plutonium deposited in a wound or in a lung, is quickly and strongly bound to transferrin, the iron transport protein found in the plasma of mammals. Small amounts of plutonium are associated with other macroglobulins or complexed with low molecular-weight substances such as citrate, sugars and peptides [32,43,51−54]. While the exact nature of the binding of Pu(IV) to transferrin is unknown, it appears to be bound by the same sites that bind iron. As with iron, bicarbonate is required in the formation of the Pu-transferrin complex [55]. Plutonium is displaced by Fe(III) and does not bind to iron saturated transferrin [51,55]. Titrimetric experiments show that transferrin specifically binds Th(IV) at at the same sites as Fe(III) [56]. Further, the half-life for the removal of plutonium from circulation nearly equals that of iron, such that after 1 h 70% of the injected plutonium is still in circulation. In contrast, 86% of the injected Am(III) or Cm(III) is removed from the blood within 1 min. Thus, the trivalent actinides are not complexed by transferrin, but are weakly associated among various plasma proteins [53,54]. The complexation of plutonium by transferrin effectively prevents its excretion, but small amounts are excreted as the citrate complex in the urine [51].

Colloids and particles of insoluble plutonium compounds which enter the circulatory system are not complexed by transferrin, but accumulate primarily in the liver. Small amounts are also found in the spleen and bone marrow. These organs have a high concentration of reticuloendothelial cells, which act as filters to consume rapidly any colloidal particles [54,57−59]. While the extent of hydrolysis depends on the oxidation state, a portion of an intravenously injected, hydrolyzable salt of plutonium, such as the nitrate or the chloride, forms insoluble colloids of hydrolyzed plutonium that are removed mainly by the liver. The remainder is complexed and transported by transferrin [10,54].

Circulating as the Pu-transferrin complex, plutonium is initially distributed throughout the body, but is eventually deposited as single atoms primarily on bone surfaces close to the marrow and the circulatory system [10]. Initially the plutonium appears to bind to the glycoproteins present in the organic matrix of bone. These proteins contain many free carboxyl groups and bind plutonium stronger than a 30-fold excess of bone mineral or any other protein investigated, including transferrin [32,51,60]. The carboxyl groups of the proteins appear to be important in binding Pu(IV), but not Am(III) or Cm(III), which are less strongly bound. The trivalent actinides are uniformly distributed on all bone surfaces and tend to deposit on bone mineral to a greater extent than plutonium [10,43].

Once deposited on bone, plutonium is not released until the bone is physically destroyed. It may become buried under a new layer of mineral or may be taken up by special cells that digest foreign materials. As these cells die, the plutonium accumulates in immobilized deposits of hemosiderin, an insoluble iron storage protein that contains a large core of polymeric iron hydroxides and phosphates. These deposits are located close to the bone surfaces in the reticuloendothelial cells of the bone marrow [10].

Table 2 Schematic structures of some chelating agents used in plutonium therapy

Ascorbic acid	![Ascorbic acid structure] O=○, HO, OH, -CHCH$_2$OH, OH, H
BAETA	HOOCCH$_2$, HOOCCH$_2$ N(CH$_2$)$_2$O(CH$_2$)$_2$N, CH$_2$COOH, CH$_2$COOH
BAL	SH; HOCH$_2$CHCH$_2$SH
Benzohydroxamic acid	O, OH; C—NH
N, N-Bis(2-hydroxyethyl)glycine	HOOCCH$_2$N, CH$_2$CH$_2$OH, CH$_2$CH$_2$OH
Catechol	OH, OH
Citric acid	CH$_2$COOH; HO—C—COOH; CH$_2$COOH
Creatine	NH$_2$; HN=C—NCH$_2$COOH; CH$_3$
Cysteine	HSCH$_2$CHCOOH; NH$_2$
2, 3-Dihydroxybenzoylglycine	HO, OH, O; C—NHCH$_2$COOH

Table 2   (Continued)

DiMeCAMS	
DTPA	HOOCCH₂ and CH₂COOHCH₂COOH structure: $N(CH_2)_2N(CH_2)_2N$ with HOOCCH₂ and CH₂COOH groups
EDTA	HOOCCH₂ and CH₂COOH structure: $N(CH_2)_2N$ with HOOCCH₂ and CH₂COOH groups
N, N′-Ethylene bis[N-phosphono-methyl]glycine	HOOCCH₂, H₂O₃PCH₂ / $N(CH_2)_2N$ \ CH₂COOH, CH₂PO₂H₂
Lactic acid	$CH_3CHCOOH$ with OH
Methionine	$CH_3SCH_2CH_2CHCOOH$ with NH₂
Neoaspergillic acid	structure
NTA	$HOOCCH_2N$ with CH₂COOH and CH₂COOH
Nicotinic acid	structure
Picolinic acid	structure

Table 2 (Continued)

| 2, 3-Pyridinedicarboxylic acid | 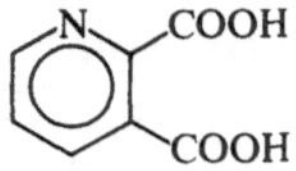 |

2, 6-Pyridinedicarboxylic acid

Pyruvic acid

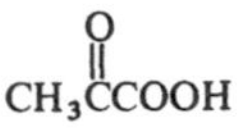

$$CH_3\overset{\overset{\textstyle O}{\|}}{C}COOH$$

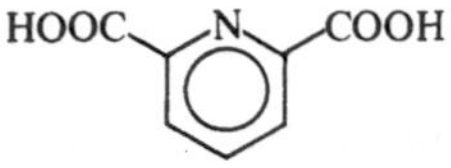

Rhodotorulic acid

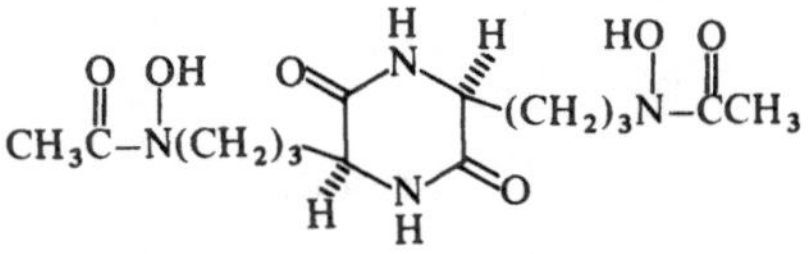

Salicylic acid

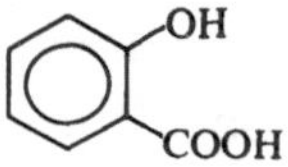

Tartaric acid

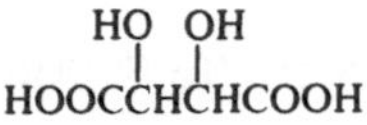

$$HOOC\overset{\overset{\textstyle HO}{|}}{C}H\overset{\overset{\textstyle OH}{|}}{C}HCOOH$$

TAAHA

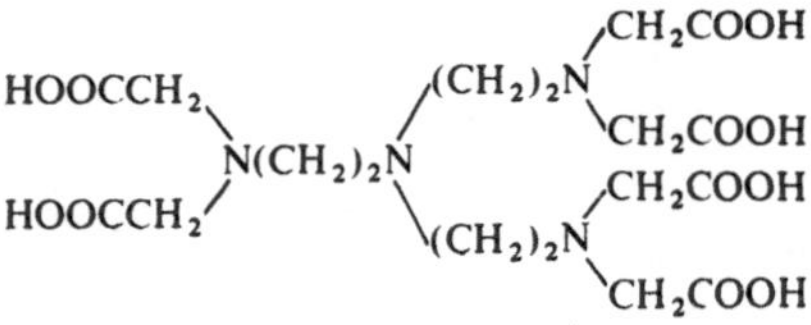

TPHA

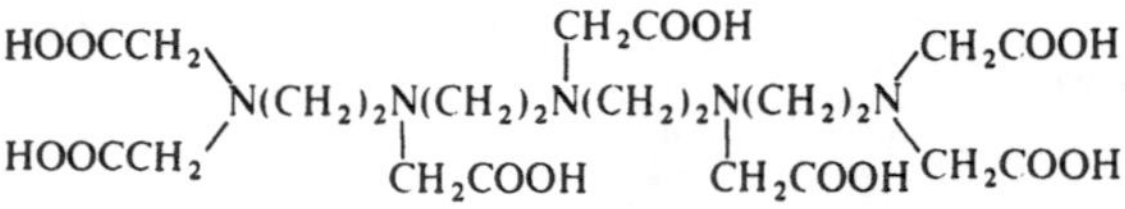

TTHA

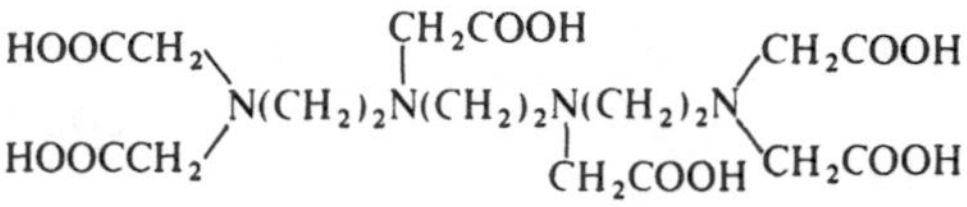

In addition to deposition on bones, smaller, but significant amounts of circulating plutonium is deposited in the liver[10]. Initially the plutonium is distributed throughout the liver, where it is bound principally in the cytosol of cells to an unidentified protein that has the chromatographic characterisitcs of a γ-globulin[61]. Within several days, the plutonium becomes associated with subcellular structures, where it is primarily bound to ferritin, a soluble iron storage protein. Small amounts of plutonium are associated with other proteins located on the subcellular structures such as glucose-6-phosphatase, cytochrome-c-oxidase, aryl-sulphatase, acid-phosphatase and unknown glycoproteins[51,62]. In an attempt to minimize their toxic effects, other toxic metals are similarly immobilized on subcellular structures[43].

As the liver cells die, the plutonium accumulates in the hemosiderin of the reticuloendothelial cells[10]. As in the bone marrow and the liver, plutonium in the spleen and the adrenal glands is also localized with hemosiderin[43,63]. Incorporation of plutonium into hemosiderin or the mineral matrix of bone is not permanent, but the mechanisms of release are not known. However, it is more probable that released plutonium will be complexed by transferrin and redeposited instead of excreted[10]. In fact, the human iron transport system is so efficient in preventing plutonium excretion that only 20–30% of the plutonium injected into humans was excreted during 27.4 years[12]. In view of the role of iron transport and storage proteins in the mammalian metabolism of plutonium, it is not surprising that the highest uptake of plutonium occurs in plants grown in iron deficient conditions[53].

III. Therapeutic Removal of Plutonium

1. Colloidal Scavenging Agents

One of the earliest attempts to remove plutonium from mammals was based on the premise that an innocuous metal ion with metabolic properties similar to plutonium would displace plutonium from body tissues — as occurs on an ion exchange resin. Because of its low toxicity in rodents and its rapid elimination from the body, zirconium was the most promising of the metals tested[64]. The details of the biological testing have been summarized in previous reviews[13,65,66]. Typically, 50–60% of the injected plutonium was rapidly excreted in the urine of rats which received an injection of 40–50 mg of zirconium in the form of zirconyl citrate within one hour of the plutonium administration, while only 1–2% of the injected plutonium was excreted by untreated rats[66–69]. Prompt treatment with zirconyl citrate was reported to remove up to 90% of the injected plutonium from dogs[70]. However, the amount of excreted plutonium dropped rapidly as the time between treatment and plutonium administration increased. When two hours elapsed between plutonium and zirconyl citrate injections, only 10% of the injected plutonium was excreted[64]. Treatment with zirconyl citrate 2 1/2 years after the plutonium injection in dogs increased the excretion of plutonium 10–15 fold, but the initial level of excretion was so low that the additional amount of plutonium removed was negligible[69].

These results indicate that zirconyl citrate is effective only in the removal of plutonium from the circulation system and not from body tissues. This is consistent with the reduction of plutonium in the blood of treated rats to 50% of the control value after five minutes and to 10% after 1 h [66,71]. The actual mechanism of plutonium removal probably involves the hydrolysis of zirconium to form colloidal aggregates of zirconium hydroxides and phosphates. Other hydrolyzable metals, such as plutonium and thorium, either coprecipitate with the zirconium or are absorbed by the colloids, which act as carriers [65,66]. In an analogous manner, the high affinity of plutonium (IV) for colloidal iron hydroxide probably explains the strong association of plutonium (IV) to ferritin and to iron storage pigments such as hemosiderin [63]. As predicted by this mechanism, manganese, iron, titanium, aluminum and thorium, metals which hydrolyze under physiological conditions, also serve as carriers [72]. However, not all of these metals promoted plutonium excretion. The larger colloids do not pass through the kidneys, but are filtered from the blood by organs such as the liver, spleen and bone marrow. Thus thorium and aluminum, which hydrolyze to form large polymers, prevent the deposition of plutonium on the skeleton, but cause an increase in the amount of plutonium deposited in the liver [72].

Prompt administration of polymeric phosphates have also been successful in increasing plutonium excretion from laboratory animals [73]. Hexametaphosphate was found to reduce bone absorption of plutonium by a factor of three, but this was accompanied by an increase in the liver burden of plutonium [74]. Thus, it seems likely that plutonium and polymeric phosphates form colloids that behave similarly to those formed with zirconium, except that the phosphates are more toxic. Alternatively, phosphate groups may bind to bone. Pretreatment with ethane-1-hydroxy-1,1-diphosphoric acid or dichloromethylenediphosphoric acid inhibited the mineralization and growth of bone as well as the skeletal uptake of plutonium [75].

2. Chelating Agents

The most promising approach to the removal of incorporated plutonium uses chelating agents to form soluble, excretable complexes of plutonium. Sodium citrate was the first complexing agent to be tested for plutonium removal [76]. Although plutonium is naturally excreted as the citrate complex [77], the rapid metabolism of sodium citrate and its complexes decreases its effectiveness as a chelating agent. Administration of sodium citrate within 2 h after the injection of plutonium increased urinary excretion several fold, but the increase was not sufficient to be of practical importance [64]. However, the excretion of thorium was increased from the control value of 28% to 47% of the injected thorium by treatment with sodium citrate 30 min after the injection of thorium [78].

The limited success with sodium citrate led to the trial and error testing of other chelating agents. Despite the fact that hard Lewis acids such as plutonium do not bind strongly to sulfur ligands, the success of 2,3-dimercapto-1-propanol, BAL, as an efficient chelator for arsenic [79] led to testing its ability to remove actinides. As expected on a chemical basis, excretion of plutonium was not enhanced by treatment with BAL, methionine, or cysteine [67,80]. Several other sulfur containing compounds were also found to have a negligible effect on the excretion of lanthanides [81]. Similar results were obtained for biologically occurring complexing agents, such as ascorbic acid, nicotinic acid and creatine, as well as for nitrilotriacetic acid (NTA), and picolinic acid. However, since 70% of the yttrium administered simultaneously with therapeutic doses of ethylenediaminetetraacetic acid (EDTA) was excreted from rats in 24 h [81], the use of EDTA was suggested for plutonium removal. Rats receiving plutonium followed by EDTA in the first 24 h excreted ten times the plutonium of the control group [82]. Another study showed that an injection of EDTA immediately following the plutonium increased the urinary excretion in rats from the control value of 6% to 51% of the injected plutonium [83]. As with zirconium, a large dose of EDTA administered 30 days after the plutonium did not significantly decrease the body burden of plutonium in rats [84]. Other authors have reviewed in more detail the removal of plutonium from mammals, including humans, using EDTA [6,13,65].

Further selection of chelating agents for plutonium removal has involved the ratio of the stability constants of the plutonium and calcium complexes formed with the chelating agent. Schubert suggested that since the concentration of serum calcium is much greater than that of other metals, any chelating agent capable of complexing calcium would exist as the calcium chelate in the circulation system. Thus, similar increases in plutonium removal would be achieved by either decreasing the chelating agent's affinity for calcium or by increasing its affinity for plutonium [65]. Other endogenous metals become significant only when they are complexed much more strongly than calcium. The use of the plutonium-calcium stability constant ratio to compare the relative effectiveness of possible chelating agents was extended by Catsch to include the competition for protons [85], which is very important in comparing ligands of different basicities. In addition to the equilibria between hydrogen, calcium, plutonium and the chelating agent, the hydrolysis of plutonium and the binding of plutonium to biological components are important. Although salicylic acid binds calcium very weakly ($K_{CaL} \sim 1$), its complexation of plutonium is too weak to promote excretion [72]. Thus, while minimizing the complexation of calcium, the affinity of the chelating agent for plutonium at physiological pH must remain greater than that of biological components.

The relative affinity of polyaminocarboxylic acids for calcium is decreased by replacing carboxyl groups with hydroxyl groups. Thus, N,N-bis(2-hydroxyethyl)-glycine was more effective in promoting plutonium excretion than NTA [86], and more effective than EDTA in increasing urinary excretion of cerium [82]. The substitution of phosphate groups for carboxylate groups in EDTA increases the relative affinity for lanthanides and actinides [87]. N,N'-Ethylenebis[N-phosphonomethyl]glycine removed more plutonium from rats than EDTA or trans-1,2-cyclohexanediaminetetraacetic acid [88]. However, the completely phosphorylated analogue of EDTA was less efficient than EDTA at removing plutonium, probably because of steric complications [87]. The longer bridge length of oxybis(ethylenenitrilo)tetraacetic acid, BAETA, allows the carboxylate groups to better encapsulate the plutonium and offers an additional binding site to account for its increased ability in plutonium removal compared to EDTA [89]. Replacement of the ether oxygen in BAETA with sulfur considerably decreases its ability to remove plutonium [90].

The additional carboxylic acid group present in diethylenetriaminepentaacetic acid, DTPA, relative to EDTA increases the stability of its actinide complexes, while the complexation of calcium remains nearly constant[91]. Thus, the octadentate DTPA was found to be superior to EDTA or zirconium, and slightly more effective than BAETA, in the removal of plutonium from animals[89,92–94]. Prompt administration of a single dose of DTPA caused the excretion of 89% of the injected plutonium from pigs during the following six days, compared to 3% excreted by controls[95]. DTPA injected in dogs (1/2 h) or in mice (1 h) following the plutonium promoted the excretion of 60–65% of the injected plutonium during 24 h, compared to 2% and 6% excreted in untreated dogs[96] and mice[97]. A further delay in treatment results in less plutonium removal such that only 15% of the injected plutonium was excreted by beagles during the first day following DTPA treatment given two hours after the plutonium[98].

Delayed treatment with multiple doses of DTPA removes moderate amounts of plutonium from animals. Treatment of swine on five successive days two months after plutonium contamination removed 11–19% of the plutonium[95]. The body burden of rats was reduced to 60% of the controls by treatment with DTPA administered on day 6, 8 and 11 after the plutonium injection[99]. The largest decrease of plutonium was found in the soft tissues, but skeletal removal was more difficult, and the moderate amounts removed may not significantly reduce the number of bone tumors formed[59,100,101]. Further details on the use of DTPA in removing internally deposited plutonium may be gained from other reviews[6,13,102,103].

Further increasing the number of carboxyl groups of a polyaminocarboxylic acid did not significantly increase plutonium removal. Triethylenetetraaminehexaacetic acid, TTHA, and DTPA were nearly equally efficient at plutonium removal[90,94,104–106], but TTHA was reported to be more toxic[106]. Perhaps due to the formation of multinuclear complexes, the additional increase in the number of carboxyl groups in tetraethylenepentaamineheptaacetic acid, TPHA, resulted in a chelating agent significantly poorer in plutonium removal than DTPA, but still more effective than EDTA[90,94]. Although tri(2-aminoethyl)aminehexaacetic acid, TAAHA, and TTHA each have six carboxylic acid groups, TTHA is better able to encapsulate the metal ion and removes much more thorium from rats than does TAAHA[94]. As with EDTA, the complete phosphorylation of DTPA decreases its ability to remove plutonium[90].

The stability of the calcium complex of the naturally-occurring iron sequestering agent desferrioxamine B, DFOA, is 10^3, which is seven powers of 10 less than that of DTPA[107]. Although the stability of its Fe(III) chelate is not much greater than that of DTPA, DFOA is significantly more efficient in iron decorporation, primarily due to its decreased affinity for calcium[108]. If administered within 1 hour after an injection of plutonium, DFOA is more effective than DTPA in promoting the excretion of plutonium. However, the ability of DFOA to decorporate plutonium decreases more rapidly than DTPA as elapsed time between contamination and treatment increases; DFOA treatment begun 4–7 days after contamination was ineffective. Prompt treatment with DFOA reduced bone deposition to 1/2 the amount in DTPA treated rats, while the metabolism of DFOA deposits more plutonium in the liver, and the low pH of the kidneys causes the release of more plutonium from the more basic

hydroxamic acid groups of DFOA. Combined treatment of DFOA and DTPA removed the greatest amount of plutonium, as the plutonium freed by destruction of the
Pu-DFOA complex in the liver and the kidney is recomplexed by DTPA [109-112].
Rhodotorulic acid, 2,3-dihydroxybenzoyl-N-glycine and neoaspergillic acid, also
naturally-occurring iron sequestering agents, removed less plutonium from hamsters
than did DTPA [113].

The additive effect of DTPA and DFOA has prompted studies of the plutonium
removal exhibited by other combinations of chelating agents. The simultaneous local
administration of citric acid or 2,6-pyridinedicarboxylic acid in conjunction with
DFOA or DTPA increased the amount of plutonium nitrate absorbed and excreted
from an intramuscular site compared to using DFOA or DTPA alone. Tartaric acid,
2,3-pyridinedicarboxylic acid, lactic acid or pyruvic acid had no effect when administered with DTPA or DFOA. Citric acid or 2,6-pyridinedicarboxylic acid when administered alone solubilized much of the plutonium from the intramuscular site, but the
plutonium was redeposited in other body tissues instead of excreted [112,114]. With the
hope of enhancing systemic plutonium removal by the formation of mixed ligand
complexes, catechol, salicylic acid and benzohydroxamic acid were administered
simultaneously with DTPA, but the amount of plutonium removed did not increase [115]. The extraordinary synergistic effect originally claimed for these compounds has been refuted by the author [116].

While DTPA is currently the reagent of choice in reducing the body burden of
actinides [6], it is most effective in removing monomeric plutonium from the circulation system — thus preventing the deposition of plutonium in body tissues — which
requires prompt treatment. DTPA removes very little hydrolyzed thorium or plutonium colloids or polymers [117]. The decreasing efficacy of plutonium removal as the
time between contamination and treatment increases indicates that the plutonium
deposited in intracellular sites is unavailable for complexation. Metabolic experiments
show that intravenously injected EDTA or DTPA mix rapidly with extracellular fluid,
but are unable to cross cell walls [118-120].

Very few cases of accidental plutonium contamination are likely to create a high
blood level of plutonium. Only very small amounts of plutonium compounds are
absorbed from the gastrointestinal tract. A maximum of 2% of ingested Pu(VI)
citrate or 0.03% Pu(IV) citrate, and much less of most other compounds, was absorbed by rats [121]. This absorption was reduced by a factor of 10 by the oral administration of ion exchange resin [122]. Only a small amount of an intramuscular deposit
of plutonium nitrate was removed by an intravenous injection of DTPA, while a local
application of DTPA 1 h after contamination removed 80—90% of the plutonium.
However, much less was removed by a local DTPA treatment applied 21 days after
contamination, during which time the plutonium had formed insoluble, polymeric
hydroxides [114]. DTPA is totally ineffective in removing insoluble plutonium compounds as PuO_2 from intramuscular sites or from lungs. These conditions are best
treated by surgical excision of contaminated tissue, lung lavage, or other methods of
direct physical removal [103,123].

Protracted DTPA therapy removes plutonium as it is liberated from cells by natural processes or solubilized by body fluids from intramuscular or lung deposits. The
slowness of these processes requires DTPA administration over long periods of time to

remove significant quantities of plutonium. The usefulness of such therapy may be of little value in preventing cancer caused by the plutonium. Thus, there has been much emphasis applied to the development of a lipophillic chelating agent. The pentaethyl ester of DTPA surpassed DTPA at removing intracellular plutonium from the liver, but its enhanced toxicity prevents its use as a therapeutic agent [124]. Several mono-amides and monoesters of EDTA and DTPA were formed using long chain alkyl amines and alcohols, but none of these derivatives removed intracellular plutonium. However, the same group has reported a derivative of DTPA that removes significant quantities of plutonium from the liver of hamsters [113,120]. A lipophilic derivative of DFOA, N-stearoyldesferrioxamine, was also tested but it was no better than DTPA at removing intracellular plutonium [113]. Two chelating antibiotics, vancomycin and cephalothin, were also ineffective at plutonium removal, either alone or in conjunction with DTPA [125].

The reticuloendothelial cells, which are especially concentrated in liver, spleen, lung, and bone marrow, rapidly remove colloids and polymers from the circulation system. Thus, it was hoped that an EDTA-cysteine copolymer would be phagocytized and release EDTA after degradation of the polymer within the cell. However, administration of the EDTA-cysteine copolymer did not increase the elimination of intracellular plutonium from rats [113]. The administration of ^{14}C-EDTA encapsulated in lipid spherules, liposomes, resulted in a high intracellular concentration of chelating agent, such that 42% of the EDTA was distributed in the liver cells of mice. When administered three days after the plutonium, liposome encapsulated DTPA removed significantly more plutonium from the liver and the skeleton than did nonencapsulated DTPA. However, the majority of the plutonium was not removed. Although the amount of plutonium removed decreased, the relative advantage of the encapsulated form increased with an increasing delay in therapy [126,127]. Encapsulation of DTPA did not increase its effectiveness when administered one day after the plutonium in rats and hamsters [128].

Glucan, a polysaccharide found in the cell walls of yeast, is also removed by the reticuloendothelial cells and its administration increases the amount of plutonium stored in the liver that is available for complexation and removal by DTPA. It is hypothesized that glucan and plutonium are associated with lysosomes, the subcellular organells that are responsible for the digestion of foreign materials. The glucan is partially hydrolyzed, which results in the osmotic swelling and dispersion of the polysaccharide and the stored plutonium [129-131]. Similarly, up to 50% of the plutonium remaining in the liver after DTPA treatment has been removed by treatment with DTPA and copolymers of divinyl ether and maleic anhydride or of acrylic and itaconic acids [132,133]. Neither glucan nor the synthetic polymers promote plutonium removal from bone and they increase the amount of plutonium in the spleen. In a similar manner, an additional 10% of americium was removed from the skeleton when an osteoporotic agent, which etches the bone surface, was used in conjunction with DTPA [134].

Despite its ability to remove much of the soluble plutonium present in body fluids, DTPA is not an exceptional chelating agent for tetravalent actinides. The formation constant of its plutonium complex is too low to displace hydroxides from the colloids and polymers of hydrolyzed plutonium or solubilize compounds such as PuO_2 at physiological pH. In addition, the inability of DTPA to completely coordinate the tetravalent actinides is shown by the easy formation of ternary complexes between Th(DTPA) and many bidentate ligands[135-137]. The hydrolysis of Th(IV) and U(IV) DTPA complexes at pH near 8 is explained by the dissociation of H^+ from a coordinated water molecule[138-141]. Further, the polyaminocarboxylic acids are toxic due to the indiscriminate complexation and removal of many metals of biological importance, primarily calcium and zinc[142-147]. While use of $CaNa_3DTPA$ prevents hypocalcemia, prolonged therapy must be frequently interrupted to allow the replenishment of other essential metal ions[147,148]. The zinc salt of DTPA is less toxic, but the larger stability constant of the Zn-DTPA complex decreases the amount of plutonium removal. However, as the time between treatment and contamination increases the difference in the amount of plutonium removed by a single dose of either salt becomes insignificant. The lower toxicity of Zn-DTPA allows larger, more frequent doses, which may remove more plutonium during extended therapy than non-toxic amounts of Ca-DTPA[85,95,98,112,123]. As exemplified by its pentaethyl ester, the toxicity of DTPA is increased by mobilization into cells where it can complex metal ions which are needed for cell functions. This casts serious doubts on the usefulness of DTPA derivatives to remove intracellular plutonium. *Thus there is a need for the development of powerful chelating agents highly specific for tetravalent actinides, particularly Pu(IV).*

IV. Synthetic Sequestering Agents Specific for Pu(IV)

Based on the similarities in the chemical and the biological transport and distribution properties of Pu(IV) and Fe(III) and the observation that microbes produce specific sequestering agents for Fe(III) that incorporate chelating groups such as hydroxamic acids and catechol, a series of sulfonated catechoylamide sequestering agents has been designed and synthesized for the specific role of complexing plutonium and other actinide(IV) ions. These synthetic macrochelates have been designed such that the chelating groups can form a cavity that gives eight-coordination about the metal and the dodecahedral geometry observed in the unconstrained actinide complexes composed of monomeric catechol ligands. The resulting compounds appear to bind tetravalent actinides strongly, while only weak complexation of trivalent and divalent metals has been observed.

It is remarkable that there are many similarities between Pu(IV) and Fe(III) (Table 3). In fact, this similarity explains much of the biological hazard posed by plutonium, as described in the previous sections of this paper. These similarities range from the charge to ionic-radius ratios for Fe(III) and Pu(IV) (4.6 and 4.2 e/Å respectively), and their formation of highly insoluble hydroxides, to their similar transport properties in mammals. These similarities of Pu(IV) and Fe(III) suggested to us a biomimetic approach to the design of Pu(IV) sequestering agents modeled after the very efficient and highly specific iron sequestering agents, siderophores, that are produced by bacteria and other microorganisms to obtain Fe(III) from the environment[150-152].

Table 3. Similarities of Pu^{4+} and Fe^{3+}

1) $\dfrac{\text{Charge}}{\text{Ionic radius}^a}$ $Pu^{4+}; \dfrac{4}{0.96} = 4.2$ $Fe^{3+}; \dfrac{3}{0.65} = 4.6$

2) $Fe(OH_3) \rightarrow Fe^{3+} + 3OH^-$ $\qquad\qquad$ $K \approx 10^{-38}$ (10^{-13} per OH^{-1})
$Fe^{3+} + H_2O \rightarrow Fe(OH)^{2+} + H^+$ $\qquad$ $K = 0.0009$

$Pu(OH_4) \rightarrow Pu^{4+} + 4OH^-$ $\qquad\qquad$ $K \approx 10^{-55}$ (10^{-14} per OH^{-1})
$Pu^{4+} + H_2O \rightarrow Pu(OH)^{3+} + H^+$ $\qquad$ $K = 0.031$ (in $HClO_4$)

3) Pu^{4+} is transported in the blood plasma of mammals as a complex of transferrin, the normal Fe^{3+} transport agent. The Pu^{4+} binds at the same site as Fe^{3+}

^{a} Ref. [149]

The siderophores (Fig. 1) typically contain hydroxamate or catecholate functional groups which are arranged to form an octahedral cavity the exact size of a ferric ion. Catechol, 2,3-dihydroxybenzene, and the hydroxamic acids, N-hydroxyamides, are very weak acids that ionize to form "hard" oxygen anions, which bind strongly to strong Lewis acids such as Fe(III) and Pu(IV). Complexation by these groups forms five-membered chelate rings, which substantially increases the stability compared to complexation by lone oxygen anions [153]. That the hydroxamic acids strongly co-ordinate tetravalent actinides is supported by the formation constants presented in Table 4. Due to its higher charge and stronger basicity, the catecholate group forms stronger complexes with the tetravalent actinides than the hydroxamic acids. Thus our goal has been the incorporation of hydroxamate or catecholate functional groups into multidentate chelating agents that specifically encapsulate tetravalent actinides.

Table 4. Formation constants for some actinide (IV) hydroxamates and catecholates

Metal	Temp, °C	$\log \beta_1{}^a$	$\log \beta_2$	$\log \beta_3$	$\log \beta_4$	Ref.
Benzohydroxamic acid, $Ph-C(O)-N(OH)-H$						
U(IV)	25	9.89	18.00	26.32	32.94	[154]
Th(IV)	25	9.60	19.81	28.76		[154]
Pu(IV)	25	12.73				[154]
N-Phenylbenzohydroxamic acid, $Ph-C(O)-N(OH)-Ph$						
Th(IV)	20				37.70	[155]
Th(IV)	25				37.80	[156]
Th(IV)	30				37.76	[157]
Pu(IV)	22	11.50	21.95	31.81	41.35	[158]
N-Phenylcinnamohydroxamic acid, $Ph-CH=CH-C(O)-N(OH)-Ph$						
Th(IV)	20	12.76	24.70	35.72	45.72	[159]
Catechol						
Th(IV)	30	17.72				[160]
4-Nitrocatechol						
Th(IV)	25	14.96	27.78	36.71	40.61	[161]

^{a} $\log \beta_n = [ML_n]/[M][L]^n$ for the reaction $M^{4+} + nL \rightarrow ML_n$ where L is the hydroxamate anion or the catecholate dianion

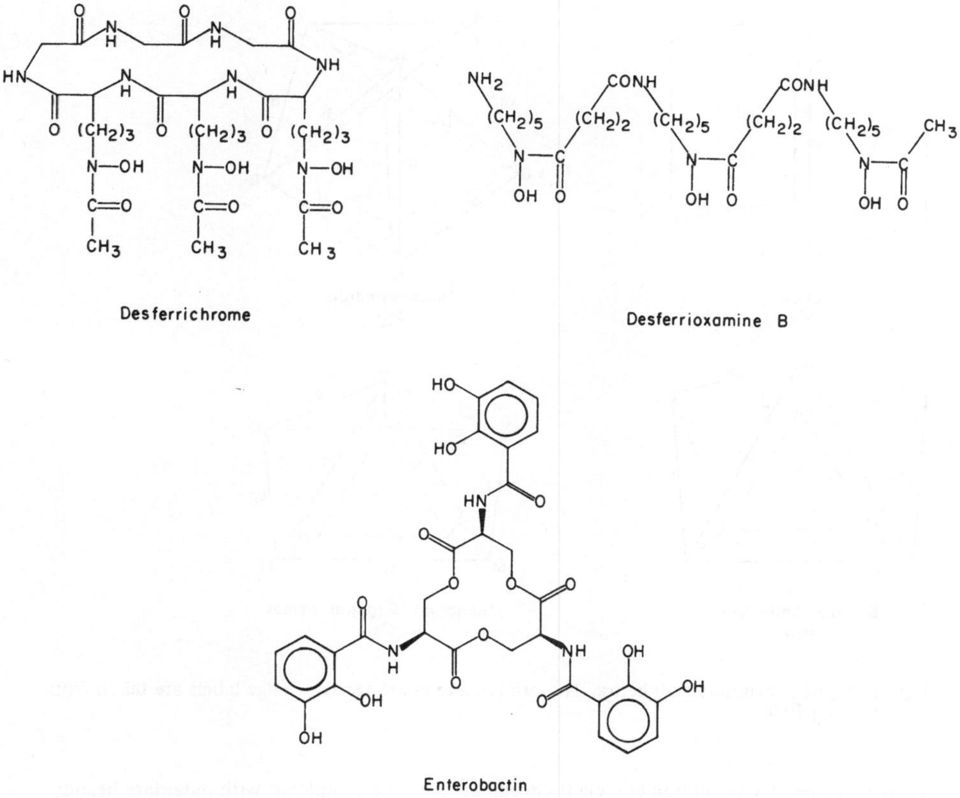

Desferrichrome

Desferrioxamine B

Enterobactin

Fig. 1. Representative siderophores

The similarity between Fe(III) and the actinide(IV) ions ends with their coordination numbers. Because of the larger ionic radii of the actinide ions, their preferred coordination number found in complexes with bidentate chelating agents is eight. Occasionally higher coordination numbers are encountered with very small ligands or by the incorporation of a solvent molecule [162,163]. Calculations of ligand-ligand repulsions indicate that either the square antiprism (D_{4d}) or the trigonal faced dodecahedron (D_{2d}) is the expected geometry for an eight coordinate complex. The coulombic energy differences between these polyhedra (Fig. 2) is very small and the preferred geometry is largely determined by steric requirements and ligand field effects. Cubic coordination lies at higher energy, but may be somewhat stabilized if f-orbital interactions were important. Another important eight coordinate polyhedron, the bicapped trigonal prism (C_{2v}), corresponds to an energy minimum along the transformation pathway between the square antiprism and the dodecahedron [164–169]. As seen in Table 5, all four of the above geometries are found in eight coordinate complexes of tetravalent actinides with bidentate ligands. However, the mmmm isomer of the trigonal faced dodecahedron is the most prevalent in the solid state.

175

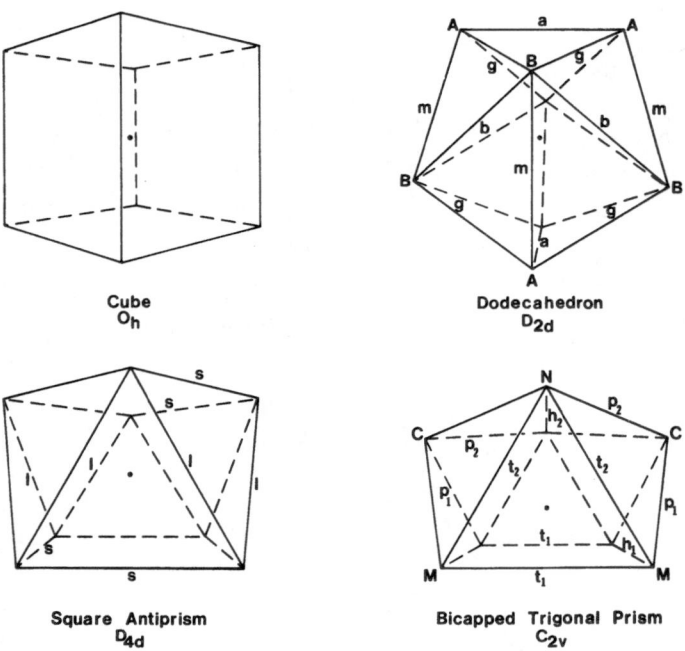

Fig. 2. Eight-coordinate polyhedra. The principal axes are vertical. Edge labels are taken from Ref. [168] and [169]

Table 5. Geometry of monomeric eight-coordinate actinide complexes with bidentate ligands

Complex [b]	Metals	Idealized Geometry [a]	Ref.
a-M(IV) (acetylacetonate)$_4$	Th, U, Ce	$h_1h_1p_2p_2$-BTP	170,171)
β-M(IV) (acetylacetonate)$_4$	Th, U, Np, Ce	ssss-SA	170,172)
M(bipyridyl)$_4$	U	ssss-Cube	173)
[M(IV) (catecholate)$_4$]$^{4+}$	Th, U, Ce	mmmm-DD	174,175)
M(IV) (dibenzoylmethanate)$_4$	Th, U, Ce	mmmm-DD	176)
M(IV) (N,N-diethyldithiocarbamate)$_4$	Th, U, Np, Pu	mmmm-DD	171,177)
[M(III) (N,N-diethyldithiocarbamate)$_4$]$^-$	Np	mmmm-DD	178)
M(IV) (diisobutrylmethanate)$_4$	U	BTP	179)
M(IV) (hexafluoroacetonylpyrazolide)$_4$	Th, U	mmmm-DD	180)
[M(III) (hexafluoroacetylacetonate)$_4$]$^-$	Am, Y, Eu	gggg-DD	181)
M(IV) (N-isopropylpivalohydroxamate)$_4$	Th	ssss-Cube	182)
M(IV) (N-isopropyl-3,3-dimethylbutano-hydroxamate)$_4$	Th	mmmm-DD	182)
M(IV) (salicylaldehydate)$_4$	Th, U	mmmm-DD	183)
M(IV) (thenoyltrifluoroacetylacetonate)$_4$	Th, U, Pu, Ce	mmmm-DD	184,185)

[a] BTP = bicapped trigonal prism, DD = trigonal faced dodecahedron, SA = square antiprism. The isomer notation is taken from references [168] and [169] and corresponds to the edges labelled in Fig. 2

[b] Thorium (trifluoroacetylacetonate)$_4$ was originally described as a llll-SA (Ref. [186]), but a reinvestigation established the presence of a coordinated water molecule forming a nine-coordinate complex (Ref. [187])

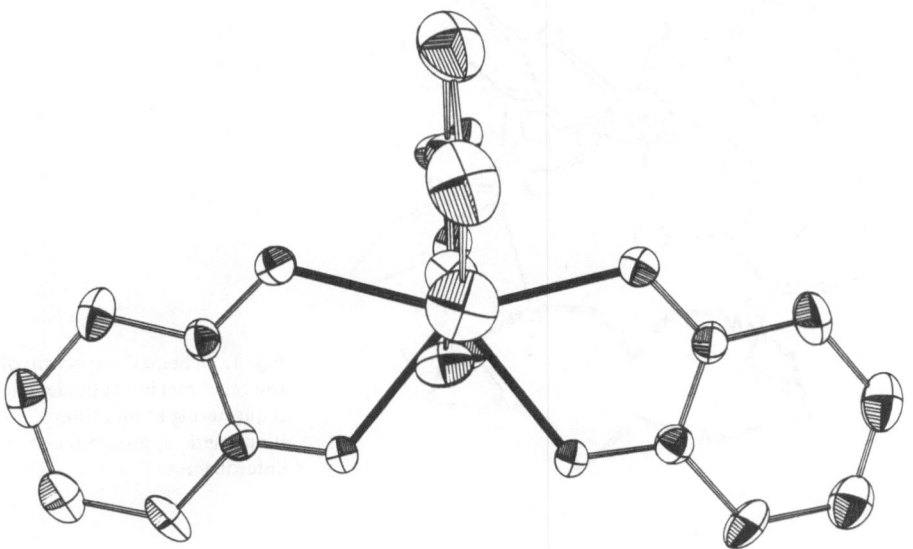

Fig. 3. The [M(catechol)$_4$]$^{4-}$ anion (M = Hf, Ce, Th, U) viewed along the mirror plane with the 4 axis vertical

Two fundamental questions in the design of an actinide-specific sequestering agent are the coordination number and geometry actually preferred by the metal with a given ligand. The structures determined for the actinide(IV) catecholates and hy-droxamates, in which the steric constraints of a macrochelate are absent, indicate that the mmmm-isomer of the dodecahedron (Fig. 3) is preferred. For maximum stability and specificity this geometry should be achieved by the ligating groups of an optimized sequestering agent that encapsulates the tetravalent actinide in a cavity with a radius near 2.4 Å. An examination of molecular models showed that this could be accomplished by the attachment of four 2,3-dihydroxybenzoic acid groups to the nitrogens of a series of cyclic tetraamines via amide linkages as shown schematically in Fig. 4. The size of the cavity formed is controlled by the ring size of the tetra-azacycloalkane backbone such that a 16 membered ring appeared most promising for the actinides. Two tetra-catechol chelating agents were synthesized from 2,3-dihy-droxybenzoic acid and 1,4,8,11-tetraazacyclotetradecane or 1,5,9,13-tetraazacyclo-hexadecane [188]. Subsequent biological evaluation in mice showed that these com-pounds reduced the accumulation of plutonium in bone and liver. However, the actinide complex apparently dissociated at low pH and released the plutonium in the animals' kidneys [97]. Titrimetric studies of these ligands showed that while they strongly complex tetravalent actinides, simple one-to-one complexes are not formed at or below neutral pH [189].

The performance of a ligand at low pH can be improved by increasing its acidity, thus reducing the competition with protons. The acidity of the catechol groups can be increased by the introduction of strongly electron withdrawing groups to the

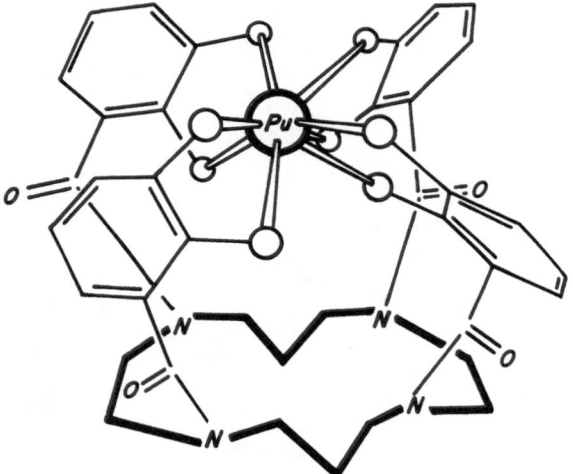

Fig. 4. Schematic structure of the tetracatechol actinide sequestering agents from a biomimetic approach based on enterobactin

aromatic rings. A more acidic analog of the above ligands was prepared from 2,3-dihydroxy-5-nitrobenzoic acid and 1,4,8,11-tetraazacyclotetradecane. The nitro groups converted the ligand into an acutely active poison and substantially changed its solubility characteristics such that a large amount of plutonium was found in the soft tissues of the treated mice[97]. In sharp contrast, sulfonation at the 5 position of each 2,3-dihydroxybenzoyl group in the ligands prepared above improved their water solubility, stability to air oxidation *and* affinity for actinide(IV) ions at low pH[190]. The increased acidity of the sulfonated derivatives prevented the deposition of plutonium in the kidneys of mice and promoted significant plutonium excretion without any appreciable toxic affects[97].

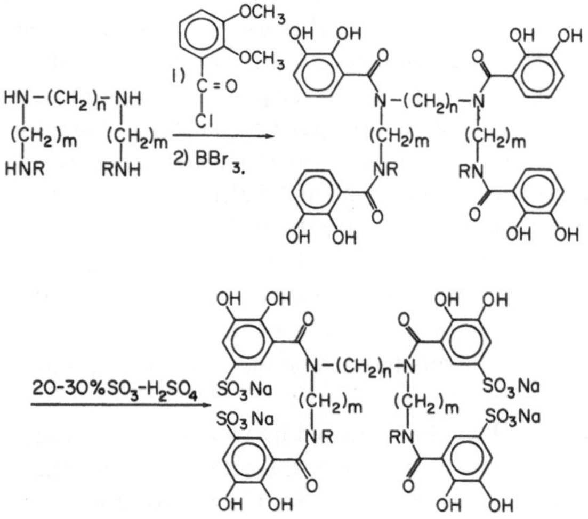

Fig. 5. General synthesis and structure of catechoylamides. The cyclic catechoylamides, in which $R = (CH_2)_p$ are abbreviated as n, m, p, m-CYCAM. The sulfonated and the analogous nitro derivatives are indicated by n, m, p, m-CYCAMS and n, m, p, m-CYCAM-NO$_2$ respectively. The linear sulfonated catechoylamides are abbreviated as m, n, m-LICAMS. A prefix is added to indicate terminal N substituents

In order to examine the effect of greater stereochemical freedom, some tetra-2,3-dihydroxy-5-sulfobenzoyl derivatives of linear tetraamines have also been prepared[190]. Maximum stability and specificity towards the actinides was obtained by optimizing the length of the methylene bridges between the amine functionalities. Butylene bridges between the nitrogens of the linear tetraamines gave better results in animal studies than ethylene or propylene bridges. The linear derivatives are significantly more effective than the cyclic catechoylamides in removing plutonium from mice[97]. In accordance with the trans configuration of amine hydrogens found in the structure of 1,5,9,13-tetraazacyclohexadecane[191], adjacent catechoylamide groups are expected to lie on opposite sides of the macrocycle. While inversion about amides is well known, it may not be facile enough in these compounds for ready coordination of the actinide by all four catechol groups.

The sulfonated catechoylamide derivatives of linear tetraamines (Table 6) are the most promising actinide sequestering agents yet prepared. The 4,4,4- or 3,4,3-LICAMS were the most efficient of the catechoylamides tested. A single dose of either ligand administered one hour after the plutonium eliminated about 65% of the injected plutonium from mice[97]. Perhaps more significant is the fact that in addition to sequestering the plutonium from body fluids, skeletal plutonium was reduced to 22% of the control value at the time of ligand injection by 3,4,3-LICAMS. Monomeric N,N'-dimethyl-2,3-dihydroxy-5-sulfobenzamide, DiMeCAMS, and 2,3-dihydroxy-benzoic acid removed very little if any plutonium and similar results were obtained for 2,3-dihydroxybenzoyl-N-glycine by Bulman and coworkers[113]. This dramatic difference between the monomeric catechols and the synthetic tetracatechol compounds confirm our original design concept that a macrochelate would be effective biologically in Pu(IV) removal. Of the sulfonated catechoylamides only the 4,4,4-LICAMS showed any toxic effects in mice.

For comparison, DTPA, the most effective conventional chelating agent, was examined and found to remove 63% of the injected plutonium. However, the dose-response curve, Fig. 6, shows that 3,4,3-LICAMS is much more effective than DTPA at lower doses — up to a two order of magnitude difference[192]. This is a good indication that endogenous metals are not strongly bound by 3,4,3-LICAMS, while metals

Table 6. Summary of actinide sequestering properties of tetrameric catechoylamides

Cyclic	
3,3,3,3-CYCAM	Mobilizes Pu but deposits it in kidneys
3,2,3,2-CYCAM–NO$_2$	Very toxic
3,3,3,3-CYCAMS $\rbrace$	Sulfonation increases acidity and solubility,
2,3,3,3-CYCAMS	prevents Pu deposition in kidneys
Linear	
2,3,2-LICAMS	Least effective of linear compounds
3,3,3-LICAMS $\rbrace$	Longer chain length, slight improvement,
4,3,3-LICAMS	still not very effective
4,4,4-LICAMS	Slightly toxic $\quad$ $\rbrace$ Longer central bridge
3,4,3-LICAMS	Derivative of spermine $\rbrace$ gives optimum geometry
	(a natural product)

Less constrained linear structures are superior to corresponding cyclic compounds

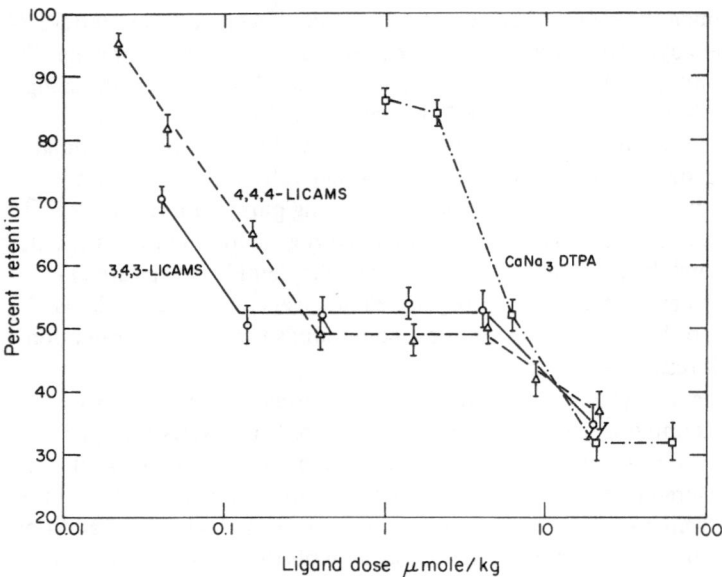

Fig. 6. Dose response comparison between LICAMS and CaNa$_3$DTPA for ^{238}Pu removal in mice

such as calcium and zinc bind strongly to DTPA, reducing the effective concentration of ligand available to complex plutonium. Thus a much larger amount of DTPA is required to achieve the same effect of a smaller quantity of 3,4,3-LICAMS, because of both a lower intrinsic affinity for actinide(IV) ions as well as a lower specificity. Of all the sequestering agents tested to date, 3,4,3-LICAMS, a derivative of the natural product spermine, is the most effective in plutonium removal at low dosages.

The greater efficacy of plutonium decorporation by 3,4,3-LICAMS compared to DTPA has also been observed in beagles[193]. A single intravenous injection of 3,4,3-LICAMS administered 30 min after the actinides removed about 86% of the injected plutonium, much better than the 70% removed by DTPA. Treatment with a combined dose of 3,4,3-LICAMS and DTPA removed very slightly more plutonium than 3,4,3-LICAMS alone. Serious toxic effects were seen in the kidneys of all dogs treated with 3,4,3-LICAMS, although the dose response curve of Fig. 6 suggests that smaller doses should be nearly as effective and would avoid such toxic effects. In contrast, DTPA is much more effective in americium decorporation. This was expected since the affinity of catechol ligands for the larger and less acidic Ln(III) or An(III) ions is quite low. The measured ratio of the tetrakis(catecholato)Ce(IV)/Ce(III) formation constants of 10^{36} is an indication of the decreased relative affinity of 3,4,3-LICAMS for the trivalent versus tetravalent actinides[175].

V. Summary

We have briefly reviewed the biological hazards associated with the actinide elements. The most abundant transuranium element produced by both industrial nuclear power plants and nuclear weapons programs is plutonium. It is also potentially the most toxic — particularly due to its long-term hazard as a carcinogen if it is introduced into the body. This toxicity is due in large part to the chemical and biochemical similarities of Pu(IV) and Fe(III). Thus in mammals plutonium is transported and stored by the transport and storage systems for iron. This results in the concentration and long-term retention of an alpha emitting radionuclide (^{239}Pu) at sites such as the bone marrow where cell division occurs at a high rate.

The earliest attempts at removal of actinide contamination by chelation therapy were essentially heuristic in that sequestering agents known to be effective at binding other elements were tried with plutonium. The research described here is intended to be a rational approach that begins with the observation that since Fe(III) and Pu(IV) are so similar, and since microbes produce agents called siderophores that are extremely effective and selective sequestering agents for Fe(III), the construction of similar chelating agents for the actinides should be possible using the same chelating groups found in the siderophores. The incorporation of four such groups (primarily catechol and hydroxamic acid) results in multidentate chelating agents that can completely encapsulate the central actinide(IV) ion and achieve the eight coordinate environment most favored by such ions. The continuing development and improvement of such sequestering agents has produced compounds which remove significant amounts of plutonium deposited in bone and which remove a greater fraction of the total body burden than any other chelation therapy developed to date.

Acknowledgments. We wish to acknowledge and thank Dr. F.L. Weitl, who has ably and creatively synthesized the macrochelating agents described here, and Dr. P.W. Durbin and her co-worker S. Jones, whose evaluation of these compounds in mice has been essential in the modification and improvement of these materials. This research is supported by the Division of Nuclear Sciences, Office of Basic Energy Sciences, U.S. Department of Energy under Contract No. W-7405-Eng-48.

VI. References

1. Blomeke, J.O., Nichols, J.P., Mclain, W.C.: Phys. Today *26*, 36, Aug. 1973
2. Kube, A.S., Rose, D.J.: Science *182*, 1205 (1973)
3. Bienvenu, P., Nofre, C., Cier, A.: C.R. Acad. Sci. *256*, 1043 (1963)
4. Finkle, R.D. et al.: Chicago Univ. Metall. Lab., Rep. CH-3783 (MDDC-1140), 1946
5. Stannard, J.N.: In: The Health Eff. of Plutonium and Radium, Proc. Sym., Sun Valley, Idaho, 1975 (Jee, W.S.S., ed.), pp. 363–372. Salt Lake City: J.W. Press 1976
6. Durbin, P.W.: Handb. Exp. Pharmacol. *36*, 739 (1973)

7. Denham, D.H.: Health Phys. *16*, 475 (1969)
8. Bair, J.C., Thompson, R.C.: Science *183*, 715 (1974)
9. Jones, M.M., Pratt, T.H.: J. Chem. Ed. *53*, 342 (1976)
10. Durbin, P.W.: Health Phys. *29*, 495 (1975)
11. Int. Comm. Radiol. Prot., Publ. 19: The Metabolism of Compounds of Plutonium and Other Actinides, Oxford: Pergamon Press 1972
12. Rundo, J. et al.: In: Diagn. Treat. Inc. Radionuclides, Proc. Int. Semin., Vienna, 1975. Int. At. Energy Agency, 1976, pp. 15−23
13. Vaughan, J., Bleany, B., Taylor, D.M.: Handb. Exp. Pharmacol. *36*, 349 (1973)
14. Moskalev, Yu. I. et al.: In: Raspred. Biol. Deistive Radioakt. Izot. (Moskalev, Yu. I. ed.), Moscow, 1966, pp. 346−365; USAEC Rep. AEC-tr-6944 (Rev) (Engl. Transl.), 1966, pp. 441−462
15. Moskalev, Yu. I.: Health Phys. *22*, 723 (1972)
16. Finkel, M.P., Biskis, B.O.: Health Phys. *8*, 565 (1962)
17. Jee, W.S.S.: In: Pathology of Irradiation (C.C. Berdjis, ed.), p. 377. Baltimore; Williams and Wilkins 1971
18. Bair, W. J., Richmond, C.R., Wacholz, B.W.: USAEC Rep. WASH-1320, 1974
19. Stannard, J.N.: Handb. Exp. Pharmacol. *36*, 669 (1973)
20. Bulman, R.A.: Struct. and Bonding *34*, 39 (1978)
21. Keller, V.C., de Alleluia, I.B.: Chemiker Z. *103*, 139 (1979)
22. Bulman, R.A.: Natl. Radiol. Prot. Board, Rep. NRPB-R44, 1976
23. Romney, E.M., Davis, J.J.: Health Phys. *22*, 551 (1972)
24. Bair, W. J.: Adv. Radiat. Biol. *4*, 255 (1974)
25. Bair, W. J.: In: Diagn. Treat. Inc. Radionuclides, Proc. Int. Semin., Vienna, 1975, Int. At. Energy Agency, 1976, pp. 51−83
26. Ross, D.M.: In: Diagn. Treat. Deposited Radionuclides, Proc. Symp., Richland, Wash., 1967 (Kornberg, H.A., Norwood, W.D. eds.), p. 427. Amsterdam: Excerpta Media Foundation 1968
27. Bair, W. J. et al.: Handb. Exp. Pharmacol. *36*, 503 (1973)
28. Markley, J.F., Rosenthal, M.W., Lindenbaum, A.: Int. J. Radiat. Biol. *8*, 271 (1964)
29. Johnson, L. J. et al.: In: The Radiobiology of Plutonium (Stover, B. J., Jee, W.S.S. eds.), pp. 213−220. Salt Lake City: J.W. Press 1972
30. Stradling, G.N. et al.: Health Phys. *35*, 229 (1978)
31. Nénot, J.C. et al.: Health Phys. *22*, 657 (1972)
32. Taylor, D.M.: Handb. Exp. Pharmacol. *36*, 323 (1973)
33. Kraus, K.A.: In: Proc. Int. Conf. Peaceful Uses At. Energy, 1st, Geneva, 1955, New York: United Nations Publ., 1956, Vol. 7, pp. 245−257
34. Ockenden, D.W., Welch, G.A.: J. Chem. Soc. 3358 (1956)
35. Grebenshchikova, V.I., Davydov, Yu. P.: Sov. Radiochem. (Engl. Transl.) *3* (2), 155 (1961); Radiokhimija *3*, 155 (1961)
36. Lindenbaum, A., Westfall, W.: Int. J. Appl. Radiat. Isot. *16*, 545 (1965)
37. Costanzo, D.A., Biggers, R.E.: Oak Ridge Natl. Lab., Rep. ORNL-TM-585, 1963
38. Starik, I. Ye., Ginzberg, F.L.: Sov. Radiochem. (Engl. Transl.) *1*, 215 (1960); Radiokhimiya *1*, 435 (1960)
39. Katz, J. J., Seaborg, G.T.: The Chemistry of the Actinide Elements, New York: Methuen 1957
40. Scott, K.G. et al.: J. Biol. Chem. *176*, 283 (1948)
41. Kendysh, I.N.: In: Raspred. Biol. Deistive Radioakt. Izot. Moskalev (Yu. I. ed.), Moscow, 1966, pp. 22−29; USAEC Rep. AEC-tr-6944 (Rev) (Engl. Transl.), 1966, pp. 25−34
42. Maletskos, C. J. et al.: In: Delayed Eff. of Bone-Seeking Radionuclides, Proc. Int. Symp., Sun Valley, Idaho, 1967 (Mays, C.W. ed.), p. 29−49. Salt Lake City: Univ. of Utah Press 1969
43. Taylor, D.M.: Health Phys. *22*, 575 (1972)
44. Yuile, C. L.: Handb. Exp. Pharmacol. *36*, 165 (1973)
45. Csevari, S., Lichner, G.: Med. Radiol. 53 (1968)

46. Jackson, S.: In: Asses. Radioact. Man, Proc. Symp., Heidelberg, 1964, Vienna, Int. At. Energy Agency, 1964, Vol. 2, p. 549
47. Milyukova, M.S. et al.: Analytical Chemistry of Plutonium, Ann Arbor: Ann Arbor-Humphrey Science 1969
48. Cleveland, J.M.: Coord. Chem. Rev. *5*, 101 (1970)
49. Jones, A.D., Choppin, G.R.: Actinides Rev. *1*, 311 (1969)
50. Degischer, G., Choppin, G.R.: In: Gmelin Handb. Anorg. Chem., 8th ed., Vol. 20, Part D1, pp. 129–176. Berlin, Heidelberg, New York: Springer 1975
51. Popplewell, D.S.: In: Diagn. Treat. Inc. Radionuclides, Proc. Int. Semin., Vienna, 1975, Int. At. Energy Agency, 1976, pp. 25–34
52. Stover, B.J., Bruenger, F.W., Stevens, W.: Radiat. Res. *33*, 381 (1968)
53. Popplewell, D.S., Boocock, G.: In: Diagn. Treat. Deposited Radionuclides, Proc. Symp., Richland, Wash., 1967 (Kornberg, H.A., Norwood, W.D. eds.), pp. 45–55. Amsterdam, Excerpta Media Foundation 1968
54. Turner, G.A., Taylor, D.M.: Phys. Med. Biol. *13*, 535 (1968)
55. Stover, B.J., Stevens, W., Bruenger, F.W. In: The Radiobiology of Plutonium (Stover, B.J., Jee, W.S.S. eds.), pp. 129–140. Salt Lake City: J.W. Press 1972
56. Harris, W.R., Carrano, C.J., Pecoraro, V.L., Raymond, K.N.: manuscript in preparation
57. Taylor, G.N. et al.: In: The Radiobiology of Plutonium (Stover, B.J., Jee, W.S.S. eds.), pp. 105–128. Salt Lake City: J.W. Press 1972
58. Matsuoka, O. et al.: Health Phys. *22*, 713 (1972)
59. James, A.C., Taylor, D.M.: Health Phys. *21*, 31 (1971)
60. Vaughn, J.: In: The Radiobiology of Plutonium (Stover, B.J., Jee, W.S.S. eds.), pp. 323–332. Salt Lake City: J.W. Press 1972
61. Boocock, G.: Radiat. Res. *42*, 381 (1970)
62. Bruenger, F.W. et al.: In: The Health Eff. of Plutonium and Radium, Proc. Sym., Sun Valley, Idaho, 1975 (Jee, W.S.S. ed.), pp. 211–221. Salt Lake City: J.W. Press 1976
63. Taylor, D.M.: In: The Radiobiology of Plutonium (Stover, B.J., Jee, W.S.S. eds.), pp. 273–279. Salt Lake City: J.W. Press 1972
64. Schubert, J.: Science, *105*, 389 (1947)
65. Schubert, J.: Ann. Rev. Nucl. Sci. *5*, 369 (1955)
66. White, M.R., Schubert, J.: J. Pharmacol. Exptl. Therap. *104*, 317 (1952)
67. Kawin, B., Copp, D.H., Hamilton, J.G.: Univ. Calif., Lawrence Berkeley Lab., Rep. UCRL-812 (1950)
68. Schubert, J., White, M.R.: J. Biol. Chem. *184*, 191 (1950)
69. Schubert, J.: J. Lab. Clin. Med. *34*, 313 (1949)
70. Joffe, M.H., Temple, L.A.: Hanford Works, Rep. HW-28636, pp. 92–97 (1952)
71. Rosenthal, M.W., Schubert, J.: Argonne Natl. Lab, Rep. ANL-5332, 1954
72. Rosenthal, M.W.: Argonne Natl. Lab, Rep. ANL-5584, pp. 100–113 (1955)
73. Tregubenko, I.P., Semenov, D.I.: Tr. Inst. Biol., Akad. Nauk SSSR, Ural. Fil. 5 (1960); Chem. Abstr. *57*, 5272d (1962)
74. Semenov, D.I., Tregubenko, I.P.: Biochemistry (Engl. Transl.) *23*, 55 (1958); Biokhimiya *23*, 59 (1958)
75. Jee, W.S.S. et al.: Res. Radiobiol. 180 (1974)
76. Painter, E. et al.: Chicago Univ. Metall. Lab., Rep. CH-3858 (AECD-2042) (1946)
77. Popplewell, D.S., Stradling, G.N., Ham, G.J.: In: The Health Eff. of Plutonium and Radium, Proc. Sym., Sun Valley Idaho, 1975 (Jee, W.S.S. ed.), pp. 245–248. Salt Lake City: J.W. Press 1976
78. Schubert, J., Wallace, H.D. Jr.: J. Biol. Chem. *183*, 157 (1950)
79. Peters, R.A.: Rec. Chem. Progr. *28*, 197 (1967)
80. Kawin, B., Copp, D.H.: Proc. Soc. Exptl. Biol. Med. *84*, 576 (1953)
81. Scott, K.G., Crowley, J., Foreman, H.: Univ. Ca., Lawrence Berkeley Lab., Rep. UCRL-587 (1949)
82. Foreman, H., Hamilton, J.G.: Univ. Ca., Lawrence Berkeley Lab., Rep. UCRL-1351 (1951)
83. Hamilton, J.G., Scott, K.G.: Proc. Soc. Exptl. Biol. Med. *83*, 301 (1953)

84. Katz, J., Weeks, M.H., Oakley, W.D.: Radiat. Res. 2, 166 (1955)
85. Catsch, A., Harmuth-Hoene, A.-E.: Biochem. Pharmacol. 24, 1557 (1975)
86. Foreman, H.: J. Am. Pharm. Assoc. Sci. Ed. 52, 629 (1953)
87. Balabukha, V.S. et al.: In: Raspred. Biol. Deistive Radioakt. Izot (Moskalev, Yu. I. ed.), Moscow, 1966, pp. 462–470; USAEC Rep. AEC-tr-6944 (Rev) (Engl. Transl.), 1966, pp. 581–591
88. Belyaev, Yu. A.: In: Raspred. Biol. Deistvie, Uskor., Vyvedediya Radioakt. Izot., Sb. Rabot, 1964, pp. 338; Chem. Abstr. 62, 16608a (1965)
89. Fried J.F. et al.: Atompraxis 5, 1 (1959)
90. Belyaev, Yu A.: USAEC Rep. AEC-tr-6408 (Engl. Transl.); Radiobiologiya 4, 760 (1964)
91. Kroll, H.: Argonne Natl. Lab, Rep. ANL-5584, pp. 150–151 (1955)
92. Smith, V.H.: Nature 181, 1792 (1958)
93. Lafuma, J., Nénot, J.C., Morin, M.: Commis. Energ. At. (Fr.) Rapp. CEA-R-3519, 1968; Chem. Abstr. 70, 27457m (1969)
94. Catsch, A.: Health Phys. 8, 725 (1962)
95. Ballou, J.E., Smith, V.H.: Handord Works, Rep. HW-65500, pp. 78–81 (1959)
96. Taylor, G.N. et al.: Health Phys. 35, 201 (1978)
97. Durbin, P.W., Jones, E.S., Weitl, F.L., Raymond, K.N.: Abstract, Radiat. Res. 83, 434 (1980)
98. Lloyd, R.D et al.: Health Phys. 35, 217 (1978)
99. Seidel, A., Volf, V.: Health Phys. 22, 779 (1972)
100. Rosenthal, M.W., Lindenbaum, A.: Radiat. Res. 31, 506 (1967)
101. Rosenthal, M.W., Lindenbaum, A.: In: Delayed Eff. of Bone-Seeking Radionuclides, Proc. Int. Symp., Sun Valley, Idaho, 1967 (Mays, C.W. ed.); p. 371. Salt Lake City: Univ. of Utah Press 1969
102. Joshima, H., Matsuoka, O.: Hoken Butsuri 13, 1 (1978)
103. Smith, V.H.: Health Phys. 22, 765 (1972)
104. Catsch, A.: Strahlenschutz Forsch. Praxis 3, 183 (1963); Chem. Abstr. 64, 12527g (1966)
105. Taylor, D.M., Sowby, F.D.: Phys. Med. Biol. 7, 83 (1962)
106. Ballou, J.E.: Nature 193, 1303 (1962)
107. Anderegg G., L'Eplattenier, F., Schwarzenbach, G.: Helv. Chim. Acta 46, 1400 (1963)
108. Imhof, P.: Ciba Symp. 11, 48 (1963)
109. Smith, V.H.: Nature 204, 899 (1964)
110. Volf, V., Seidel, A., Takada, K.: Health Phys. 32, 155 (1977)
111. Taylor, D.M.: Health Phys. 13, 135 (1967)
112. Volf, V.: In: Diagn. Treat. Inc. Radionuclides. Proc. Int. Semin., Vienna, 1975, Int. At. Energy Agency, 1976, pp. 307–322
113. Bulman R.A., Griffin, R.J., Russel, A.T.: Natl. Radiol. Prot. Board, Rep. NRPB/RANDD-1, 1977, pp 87 89
114. Volf V : Health Phys. 29, 61 (1975)
115. Bulman R A., Crawley F.E.H., Geden, D.A.: Nature 281, 406 (1979)
116. Schubert, J.: Nature 281, 406 (1979); Univ. Utah, Coll. Medicine, Radiobiol. Div., Rep. COO-2969-1, 1977; Schubert, J., Derr, S.K.: Nature 275, 311 (1978)
117. Markley, J.F.: Argonne Natl. Lab, Rep. ANL-6637, 1962, pp. 5–8
118. Foreman, H., Trujillo, T.T.: J. Lab. Clin. Med. 43, 566 (1954)
119. Foreman, H.: In: Metal-Binding in Medicine (Seven, M.J., Johnson, L.A. eds.), pp. 82–94. Philadelphia: Lippincott, 1960
120. Stevens, W. et al.: Radiat. Res. 75, 397 (1978)
121. Weeks, M.H. et al.: Radiat. Res. 4, 339 (1956)
122. Belyaev, Yu A.: In: Biol. Deistive Radiat. Voprosy Raspred. Radioakt. Izot (Lebedinskii, A.V., Moskalev, Yu. I. eds.), Moscow, 1961; USAEC Rep. AEC-tr-5265 (Engl. Transl.), 1961, pp. 180–187
123. Catsch, A.: In: Diagn. Treat. Inc. Radionuclides, Proc. Int. Semin., Vienna, 1975, Int. At. Energy Agency, 1976, pp. 295–305
124. Markley, J.F.: Int. J. Radiat. Biol. 7, 405 (1963)
125. Smith, V.H.: Battelle Northwest Lab., Rep. BNWL-280, 1966, pp. 81–82

126. Rahman, Y.E., Rosenthal, M.W., Cerny, E.A.: Science *180*, 300 (1973)
127. Rosenthal, M.W. et al.: Radiat. Res. *63*, 262 (1975)
128. Stather, J.W. et al.: In: Diagn. Treat. Inc. Radionuclides, Proc. Int. Semin., Vienna, 1975, INt. At. Energy Agency, 1976, pp. 387–400
129. Rosenthal, M.W., Smoller, M., Lindenbaum, A.: In: Diagn. Treat. Deposited Radionuclides, Proc. Symp., Richland, Wash., 1967 (Kornberg, H.A., Norwood, W.D. eds.), pp. 403–412. Excerpta Media Foundation 1968
130. Rosenthal, M.W. et al.: Radiat. Res. *53*, 102 (1973)
131. Rosenthal, M.W. et al.: Radiat. Res. *63*, 253 (1975)
132. Guilmette, R.A., Lindenbaum, A.: In: The Health Eff. of Plutonium and Radium, Proc. Sym., Sun Valley Idaho, 1975 (Jee, W.S.S. ed.), pp. 223–232. Salt Lake City: J.W. Press 1976
133. Lindenbaum, A., Rosenthal, M.W., Guilmette, R.A.: In: Diagn. Treat. Inc. Radionuclides, Proc. Int. Semin., Vienna, 1975. Int. At. Energy Agency, 1976, pp. 357–372
134. Fisher, D.R. et al.: Univ. Utah, Coll. Medicine, Radiobiol. Div., Rep. COO-119–249, 1974, p. 161
135. Pachauri, O.P., Tandon, J.P.: J. Inorg. Nucl. Chem. *37*, 2321 (1975)
136. Pachauri, O.P., Tandon, J.P.: Indian J. Chem., Sect. A *15*, 57 (1977)
137. Pachauri, O.P., Tandon, J.P.: J. Gen. Chem. (Engl. Transl.) *47*, 398 (1977); Zh. Obshch. Khim. *47*, 433 (1977)
138. Carey, G.H., Martell, A.E.: J. Am. Chem. Soc. *90*, 32 (1968)
139. Bogucki, R.F., Martell, A.E.: J. Am. Chem. Soc. *80*, 4170 (1958)
140. Fried, A.R., Martell, A.E.: J. Am. Chem. Soc. *93*, 4695 (1971)
141. Grimes, J.H.: In: Diagn. Treat. Inc. Radionuclides, Proc. Int. Semin., Vienna, 1975, Int. At. Energy Agency, 1976, pp. 419–460
142. Cohen, N., Guilmette, R.: Bioinorg. Chem. *5*, 203 (1976)
143. Seven, M.J.: In: Metal-Binding in Medicine (Seven, M.J., Johnson, L.A. eds.), pp. 95–103. Philadelphia: Lippincott 1960
144. Catsch, A. et al.: In: Diagn. Treat. Deposited Radionucliedes, Proc. Symp., Richland, Wash., 1967 (Kornberg, H.A., Norwood, W.D. eds.), pp. 413–418. Amsterdam: Excerpta Media Foundation 1968
145. Foreman, H., Nigrovic, V,: In: Diagn. Treat. Deposited Radionuclides, Proc. Symp., Richland, Wash., 1967 (Kornberg, H.A., Norwood, W.D. eds.), pp. 419–423. Amsterdam: Excerpta Media Foundation 1968
146. Truhaut, R., Boudene, C., Lutz, M.: Ann. Biol. Clin. *24*, 419 (1966)
147. Planas-Bohne, F., Lohbreier, J.: In: Diagn. Treat. Inc. Radionuclides, Proc. Int. Semin., Vienna, 1975, Int. At. Energy Agency, 1976, pp. 505–515
148. Lloyd, R.D. et al.: Health Phys. *31*, 281 (1976)
149. Shannon, R.D.: Acta Cryst., Sect. A *32*, 751 (1976)
150. Microbial Irdn Metabolism (Nielands, J.B. ed.), New York: Academic Press 1974
151. Raymond, K.N.: Adv. Chem. Ser. *162*, 33 (1977)
152. Raymond, K.N., Carrano, C.J.: Acc. Chem. Res. *12*, 183 (1979)
153. Huhey, J.E.: Inorganic Chemistry: Principles of Structure and Reactivity, pp. 418–422. New York: Harper and Row 1972
154. Barocas, A. et al.: J. Inorg. Nucl. Chem. *28*, 2961 (1966)
155. Zharovskii, F.G., Ostrovskaya, M.S., Sukhomlin, R.I.: Izv. Vyssh. Ucheb. Zaved., Khim. Khim. Teknol. *10*, 989 (1967); Chem. Abstr. *69*, 3096y (1968)
156. Dyrssen, D.: Acta Chem. Scand. *10*, 353 (1956)
157. Reidel, A.: J. Radioanal. Chem. *13*, 125 (1973)
158. Chimutova, M.K., Zolotov, Yu A.: Sov. Radiochem. (Engl. Transl.) *6*, 625 (1964); Radiokhimiya *6*, 640 (1964)
159. Zharovskii, F.G., Sukhomlin, R.I., Ostrovskaya, M.S.: Russ. J. Inorg. Chem. (Engl. Transl.) *12*, 1306 (1967); Zh. Neorg. Khim. *12*, 2476 (1967)
160. Agrawal, R.P., Mehrotra, R.C.: J. Inorg, Nucl. Chem. *24*, 821 (1962)
161. Avdeef, A., Bregante, T.L., Raymond, K.N.: manuscript in preparation

162. Casellato, U., Vidali, M., Vigato, P.A.: Inorg. Chim. Acta *18*, 77 (1976)
163. Moseley, P.T.: In: MTP Int. Rev. Sci.: Inorg. Chem., Ser. Two (Bagnall, K.W., ed.). Vol. 7, pp. 65–110. London: Butterworth 1975
164. Burdett, J.K., Hoffmann, R., Fay, R.C.: Inorg. Chem. *17*, 2553 (1978)
165. Blight, D.G., Kepert, D.L.: Inorg. Chem. *11*, 1556 (1972)
166. Muetterties, E.L., Guggenberger, L.J.: J. Am. Chem. Soc. *96*, 1748 (1974)
167. Kepert, D.L.: Prog. Inorg. Chem. *24*, 179 (1978)
168. Hoard, J.L., Silverton, J.V.: Inorg. Chem. *2*, 235 (1963)
169. Porai-Koshits, M.A., Aslanov, L.A.: J. Struct. Chem. (Engl. Transl.) *13*, 244 (1972); Zh. Strukt. Khim. *13*, 266 (1972)
170. Allard, B.: J. Inorg. Nucl. Chem. *38*, 2109 (1976)
171. Steffen, W.L., Fay, R.C.: Inorg. Chem. *17*, 779 (1978)
172. Lenner, M.: Acta Cryst., Sect. B *34*, 3770 (1978)
173. Piero, G.D. et al.: Cryst. Struct. Commun. *4*, 521 (1975)
174. Sofen, S.R. et al.: J. Am. Chem. Soc. *100*, 7882–7887 (1978)
175. Sofen, S.R., Cooper, S.R., Raymond, K.N.: Inorg. Chem. *18*, 1611 (1979)
176. Wolf, L., Barnighausen, H.: Acta Cryst. *13*, 778 (1960)
177. Brown, D., Holah, D.G., Rickard, C.E.F.: J. Chem. Soc., Sect. A 423 (1970)
178. Brown, D., Holah, D.G., Rickard, C.E.F.: J. Chem. Soc., Sect. A 786 (1970)
179. Day, V.W., Fay, R.C.: Abstr. of Papers, Am. Crystal. Assoc. Summer Meeting 1976, p. 78
180. Volz, K., Zalkin, A., Templeton, D.H.: Inorg. Chem. *15*, 1827 (1976)
181. Burns, J.H., Danford, M.D.: Inorg. Chem. *8*, 1780 (1969)
182. Smith, W.L., Raymond, K.N.: J. Am. Chem. Soc., in press
183. Hill, R.J., Rickard, C.E.F.: J. Inorg. Nucl. Chem. *39*, 1593 (1977)
184. Lenner, M., Lindquist, O.: Acta Cryst., Sect. B *35*, 600 (1979)
185. Baskin, Y., Prasad, N.S.K.: J. Inorg. Nucl. Chem. *25*, 1011 (1963)
186. Wessels, G.F.S., Leipoldt, J.G., Bok, L.D.C.: Z. Anorg. Allg. Chem. *393*, 284 (1972)
187. Hambley, T.W. et al.: Austr. J. Chem. *31*, 2635 (1978)
188. Weitl, F.L. et al.: J. Am. Chem. Soc. *100*, 1170 (1978)
189. Harris, W.R., Avdeef, A., Raymond, K.N.: unpublished
190. Weitl, F.L., Raymond, K.N.: J. Am. Chem. Soc. *102*, 2289 (1980)
191. Smith, W.L., Ekstrand, J.D., Raymond, K.N.: J. Am. Chem. Soc. *100*, 3539 (1978)
192. Durbin, P.W., Jones, E.S., Raymond, K.N., Weitl, F.L.: Radiat. Res. *81*, 170 (1980)
193. Bruenger, F.W. et al.: Univ. Utah, Coll. Medicine, Radiobiol. Div., Rep. COO-119–254 (1979)

Novel Structures in Iron-Sulfur Proteins

António V. Xavier, José J. G. Moura and Isabel Moura

Centro de Química Estrutural, I.S.T., Av. Rovisco Pais, 1000 Lisboa, Portugal

and

Gray Freshwater Biological Institute, University of Minnesota, P.O. Box 100, USA-Navarre, MN 55392

The main objective of this article is to put together the information available on the novel iron-sulfur centers and relate their properties with those of the iron-sulfur containing proteins. Special effort is put on the techniques used to identify their centers and in the discussion of the oxidation-reduction potentials involved.

Table of Contents

A. Introduction

Iron-sulfur proteins represent a class of proteins which contain non-heme iron bound to sulfur ligands provided by cysteinyl residues of the polypeptide chain and, with the exception of rubredoxin type proteins, inorganic "labile" sulfur.

The study of the structure-function of these proteins has been one of the most active research fields challenging an ensemble of multidisciplinary sciences, which include Biology, Biochemistry, Chemistry and Physics.

The simple constitution of the active center, iron and sulfur, contrasts with the diversified role played by these proteins in key biological oxidation-reduction processes, such as carbon, hydrogen, sulfur and nitrogen metabolism, using a very wide range of redox potentials (+ 350 mV in photosynthetic bacteria to − 600 mV in chloroplasts).

The elucidation of their structures was particularly difficult due to the instability of the active centers. Structural determination efforts culminated about 10 years ago with the evidence of three basic structures:

[Rb] (rubredoxin); [2Fe−2S]; and [4Fe−4S] centers.

Simple iron-sulfur proteins, containing these basic structures are listed in Table 1. They are generally electron-transfer proteins mediating electron exchange between enzymatic systems, with the possible exception of hydrogenase, and aconitase, which might have catalytic activity of their own, as shall be discussed later.

Rubredoxin constitutes the simplest class of iron-sulfur proteins with one iron atom coordinated by four cysteinyl residues and containing no labile sulfur. They are present in aerobes as well as in anaerobic organisms but despite their widespread occurence their general function has not yet been determined, although the rubredoxin isolated from *Pseudomonas oleovorans* has been shown to be active in the ω-hydroxylation reaction [41] and a highly specific NADH−H⁺ rubredoxin-oxido-reductase is present in *Desulfovibrio gigas* [39].

The elucidation of the [2 Fe−2 S] centers was achieved by the conbination of different spectroscopic techniques, namely EPR, NMR, magnetic susceptibility, Mössbauer spectroscopy and various optical methods. This center has the possibility of transferring one electron per cluster between two stable oxidation states: the diamagnetic oxidized form (S = O) is converted upon one electron reduction into an EPR active form with a typical g = 1.94 feature (S = 1/2). Gibson et al. [27a] established a model involving antiferromagnetic coupling between the two iron atoms bridged by sulfur ligands and the model received further support by Mössbauer measurements [26a,54a]. Recently the structure was confirmed by X-Ray difraction studies on the ferredoxin from *Spirulina platensis* [79a].

The [4 Fe−4 S] cores have been one of the most intriguing inorganic structures involved in biological systems. Carter et al. (1977) [19] demonstrated that the same basic structure is present in the two [4 Fe−4 S] centers of the 8 Fe ferredoxin of *Peptococcus aerogenes* (E'_0 = − 400 mV) [2] and in the high potential iron protein (HiPIP) isolated from the purple photosynthetic bacterium *Chromatium vinosum*

$(E'_0 = + 350 \text{ mV})$[20]. The "three state" hypothesis[21] considers that the [4 Fe–4 S] clusters can be stabilized in three oxidation states, + 3, + 2 and + 1 (see Table 2). HiPIP is representative of the + 3 oxidation state. The center in the native form is in the + 2 state and is EPR silent. Upon one electron oxidation it becomes paramagnetic, exhibiting a fairly isotropic type EPR spectrum with g-values around 2.01. The + 2 oxidation state is representative both of the reduced HiPIP and of the oxidized bacterial type ferredoxins, and is the oxidation state in which these last proteins are isolated. A further one-electron reduction of bacterial type ferredoxins brings the cluster to a + 1 oxidation state, which gives a typical axial EPR signal at g = 1.94. The oxidation-reduction transition, which the cluster can undergo, can be denoted as [4 Fe–4 S]$^{+2(+2,+3)}$ for HiPIP and [4 Fe–4 S]$^{+2(+2,+1)}$ for bacterial type ferredoxins, using the IUB-IUPAC nomenclature[36]. Further evidence for the "three state" hypothesis comes from the following observations: 1) native HiPIP (+ 2 state) can be further reduced to a + 1 state[13] in unfolding conditions (Unpublished results by Rupp and Cammack in Cammack (1979)) showing the typical g = 1.94 type spectrum with a redox potential estimated to be approx. − 600 mV; and 2) clostridial ferredoxin could be superoxidized to the oxidation level + 3 with ferricyanide[78]. The state + 0 has not been described so far.

More complex iron-sulfur proteins have been described in the literature, where the iron-sulfur center is associated with other prosthetic groups such as flavin, heme, molybdenum, thiamine diphosphate (TDP), selenium and chlorophyll.

Table 3 describes the basic iron-sulfur structures and the oxidation states that can be found in simple proteins and how they may associate with other prosthetic groups.

While the structures of the iron-sulfur proteins were being examined, inorganic chemists were successfully synthesizing model compounds of their active centers and developing the spectroscopic techniques necessary for their characterization. In particular, Holm and co-workers succeeded in designing model compounds which mimic the [Rb], [2 Fe–2 S] and [4 Fe–4 S] clusters found in proteins. These models proved to be of great importance for the determination of the extent of the role played by the polypeptide chain in controling the physico-chemical properties of the inorganic iron-sulfur structure[30,31]. Also Orme-Johnson, Holm and co-workers developed methods to "displace" or "extrude" the [2 Fe–2 S] and [4 Fe–4 S] centers, which constitute important means of analysing the type of cores present in proteins. The technique involves the unfolding of the protein by exposure to anaerobic 80% DMSO in the presence of a thiol which displace the Fe–S core. Apoferredoxins (adrenodoxin and *B. polymixa* ferredoxin are used as specific acceptors for [2 Fe–2 S] and [4 Fe–4 S] cores respectively) are added. In refolding conditions the extruded cores are recaptured by the added standard apoproteins, giving typical ferredoxins and their low temperature EPR can be quantitated[60].

Other techniques were developed in order to identify the iron-sulfur centers using EPR in mild unfolding conditions[15] and other more sophisticated methods now use ^{19}F NMR spectroscopy[81].

Table 1. Basic structures in simples iron-Sulfur Proteins

Protein source	Redox center	$M_r \times 10^{-3}$	EPR g-values	Extinction coefficients per cluster (indicated wave length) $\times 10^{-3}$	Redox potential (mV)	References
1-2 iron no labile sulfur						
Clostridium pasteurianum rubredoxin[a]	[Rb]	6	4.3 (also 9.4 below 20 K) (ox)	10.8 (390) 8.8 (490)	− 57	(43, 63)
Desulfovibrio gigas desulforedoxin	[2Fe] distorted Rb	7.9 (2 × 3.8)	7.7, (5.7), 4.1, 1.8 (ox)	7.8 (370) 4.6 (507)	− 35	(50)
2-iron ferredoxin						
Pseudomonas putida	[2Fe−2S]⁺²(+2,+1)	12.5	2.02, 1.93, 1.93 (red)	5.0 (325) 10.0 (415) 9.6 (455)	− 240	(24, 26)
Clostridium pasteurianum	[2Fe−2S]⁺²(+2,+1)	25	2.01, 1.94, 1.92 (red)	16.6 (333) 9.2 (425) 10.2 (463)	− 300	(89, 90)
Spinach (higher plants)	[2Fe−2s]⁺²(+2,+1)	11	2.05, 1.95, 1.89 (red)	12.8 (325) 9.6 (420) 9.8 (465)	− 390	(62, 91)
3-iron ferredoxin						
D. gigas Ferredoxin II	[3Fe−xS]	24 (4 × 6)	2.02, 2.00, 1.97 (ox)	23.1 (305) 15.7 (405) 13.3 (453 shoulder)	− 130	(11, 17, 34, 35, 49, 50)
4 and 8 iron ferredoxin						
Bacillus polymyxa	[4Fe−4S]⁺²(+2,+1)	9	2.06, 1.92, 1.88 (red) small 2.01 (ox)	15.2 (400)	− 422	(92)
Clostridium pasteurianum	2 × [4Fe−4S]⁺²(+2,+1)	6	2.04, 2.00, 1.96, 1.92, 1.86 (red) small 2.01 (ox)	18.4 (305) 15.2 (390)	− 400	(53, 62)

Chromatium vinosum HiPIP	[4 Fe—4 S]$^{+2}$(+2,+3)	10	2.12, 2.04, 2.04 (ox)	32.4 (325), 20.0 (385), 18.4 (450), 16.0 (388)	+ 350	(18)
3-iron, 4-iron ferredoxin						
Azotobacter vinelandii Fd I or Fe-S protein III	[3 Fe—xS] + + [4 Fe—4 S]$^{+2}$(+2,+1)	14.5	2.01	27 (400)	− 420 + 340	(84)
D. gigas Ferredoxin I	[3 Fe, xS] + + [4 Fe—4 S]$^{+2}$(+2,+1)	18 3 × 6	2.02 (ox) 2.07, 1.94, 1.92 (red)	20.6 (300), 16.0 (405)	− 50 − 455	(11, 17)

^{a} A two-iron rubredoxin was isolated from Pseudomonas oleovorans[41] but the properties are similar to the one-iron rubredoxin

Table 2. Oxidation states of the [4 Fe−4 S] center

"Formal valence" of iron atoms	Oxidation level	Magnetic state	Typical redox transitions in proteins		Analogue model compounds
$4\,Fe^{3+}$	+4				
$3\,Fe^{3+},\ 1\,Fe^{2+}$	+3	para	$HiPIP_{ox}$	$Fd_{super\ ox}$	$[Fe_4S_4(SR)_4]^{-1}$
			$\updownarrow\ +350\,mV$	$\updownarrow$	
$2\,Fe^{3+},\ 2\,Fe^{2+}$	+2	dia	$HiPIP_{red}$ (native state)	Fd_{ox}	$[Fe_4S_4(SR)_4]^{-2}$
			$\updownarrow\ \approx -600\,mV$	$\updownarrow\ -400\,mV$	
$1\,Fe^{3+},\ 3\,Fe^{2+}$	+1	para	$HiPIP_{superred}$	Fd_{red}	$[Fe_4S_4(SR)_4]^{-3}$
$4\,Fe^{2+}$	+0				

The oxidation level is determined by adding the formal charges on the iron and labile sulfur atoms. The corresponding values for model compounds are decreased by four units due to the thiolate ligands that are taken into account. The designation of the formal valence is useful for a description of the oxidation level but in practice, due to the covalency of the iron atoms in the cluster structure, the iron atoms can not be distinguished.

Table 3. Active centers in iron-sulfur proteins

Iron-sulfur

Simple[a]	complex[b]	
[Rb] (1 Fe)	n x [Fe−S] +	
[Rb] type (2 Fe)	[Fe−S] +	flavin
$[2\,Fe-2\,S]^{+2(+2,+1)}$	[Fe−S] +	flavin + haem
[3 Fe−xS]	[Fe−S] +	haem
$\updownarrow$	[Fe−S] +	Mo
$[4\,Fe-4\,S]^{+2(+2,+3)}$	[Fe−S] +	Mo + flavin
$[4\,Fe-4\,S]^{+2(+2,+1)}$	[Fe−S] +	Mo + Se
$2 \times [4\,Fe-4\,S]^{+2(+2,+1)}$	[Fe−S] +	Fe-S-Mo nitrogenase cofactor
$[3\,Fe-xS] + [4\,Fe-4\,S]^{+2(+2,+3)}$	[Fe−S] +	TDP
$[3\,Fe-xS] + [4\,Fe-4\,S]^{+2(+2,+1)}$	[Fe−S] +	Chlorophyl

[a] Examples of these simple iron-sulfur are shown in Table 1
[b] For an extensive list of conjugated proteins see Cammack[14] and Averill and Orme-Johnson (1978)[5]

In the last two years, and particularly after August 1979, a renewed interest in the field of the iron-sulfur proteins was brought about by the findings of four novel structures in the active centers of these proteins which has enlarged the number of known basic iron-sulfur structures:

(I)   a three-iron cluster with a non-identified number of sulfur atoms[27,35], arranged as a unique center or participating in some more complex arrangement;

(II) the nitrogenase co-factor center containing one molybdenum and probably six iron atoms and labile sulfur;

(III) the P-clusters of nitrogenase, which may represent a variation of the [4 Fe–4 S] basic structure using different oxidation states and/or different ligation[61,88];

(IV) the isolation of desulforedoxin, a two-iron protein containing centers that represent a variation on the basic rubredoxin structure[52].

Numerous reviews have been published in this active field of research. Recent articles cover different aspects of the physico-chemistry and biology of the iron-sulfur proteins and enzymes: Cammack[14] compiled extensive information of simple and conjugated iron-sulfur containing proteins; Yoch and Carithers[85] related the physical and chemical nature of bacterial iron-sulfur clusters as they are found in enzymes and ferredoxins with general physiological aspects; Greenwood and Barber[28] surveyed the iron-sulfur protein field and references therein give access to past literature; Averill and Orme-Johnson[4] and Holm[30] published detailed information on the identification of iron-sulfur clusters and their spectroscopic parameters with special relevance to model compounds; Adman[1] compared structural data and Huynh and Münck[33] approached the inter-relation of EPR and Mössbauer to the study of the iron-sulfur centers. The three volumes edited by Lovenberg (1973, 1977)[42] must be referred as a fundamental source of background information on the field.

The main objective of this article is to put together the information available on the novel iron-sulfur centers and relate their properties with those of the iron-sulfur containing proteins. Special effort is put on the techniques used to identify their centers and in the discussion of the oxidation-reduction potentials involved.

B. Novel Structures

I. Rubredoxin Type Proteins

The rubredoxin isolated from *Cl.pasteurianum* has generally been used as a prototype of this class of proteins and an enormous amount of effort has been devoted to the determination of its structural features and electronic parameters of the iron containing center. Extensive X-Ray[80], Mössbauer[25], MCD[67] and EXFAS[71] measurements as well as comparative studies with model compounds[22,38] describe the active center where the iron is approximately tetrahedrically coordinated to four cysteinyl ligands, with average Fe–S bond length of 2.267 ± 0.003 A[71]. The variation in bond lengths within one complex is now known to be much smaller than was once thought.

A protein named desulforedoxin was isolated from a sulfate reducer organism, *Desulfovibrio gigas*, by Moura et al.[47] and which represents a variation on the basic rubredoxin type structure. Its spectroscopic features will be discussed in relation to those of rubredoxin isolated from *D.gigas* and *Cl.pasteurianum*.

Desulforedoxin is a non-heme iron-protein containing two iron atoms and no labile sulfur. It is isolated as a dimer (M_r = 7.900) consisting of two identical subunits of 36 amino-acid residues for which the sequence has been determined [10]. A comparison of the visible spectra of desulforedoxin and rubredoxin shows similarities between their chromophores [51] but the spectrum of desulforedoxin is not a superimposition of two rubredoxin spectra as is the case for the two-iron containing rubredoxin isolated from *Pseudomonas oleovorans* [41].

The oxidized form of the simple rubredoxin gives typical EPR spectra with signals at g = 4.3 and 9.4 (Fig. 1). The spectrum was interpreted by Blumberg [7] as due to the transition between the middle and lower Kramers doublets of the high-spin Fe(III) center, with a higher degree of rhombic distortion. As the temperature is lowered, the g = 4.3 signal decreases in magnitude as the middle Kramers doublet becomes depopulated, and the g = 9.4 signal from the ground state increases. Again, the rubredoxin isolated from *P.oleovorans* which contains two iron atoms per molecule gives a rubredoxin type EPR spectrum [41,63] indicating that the two iron sites are identical and do not interact magnetically. However, the oxidized form of desulforedoxin shows a complex EPR spectrum (Fig. 1) with principal features at g = 7.7 and g = 5.7 and broad features around g = 4.9 and g = 1.8 (an additional signal observed around g = 8.9 was shown to be due to adventitious material [52]). Decreasing the temperature (see Fig. 1) the signal at g = 7.7 increases while the signal

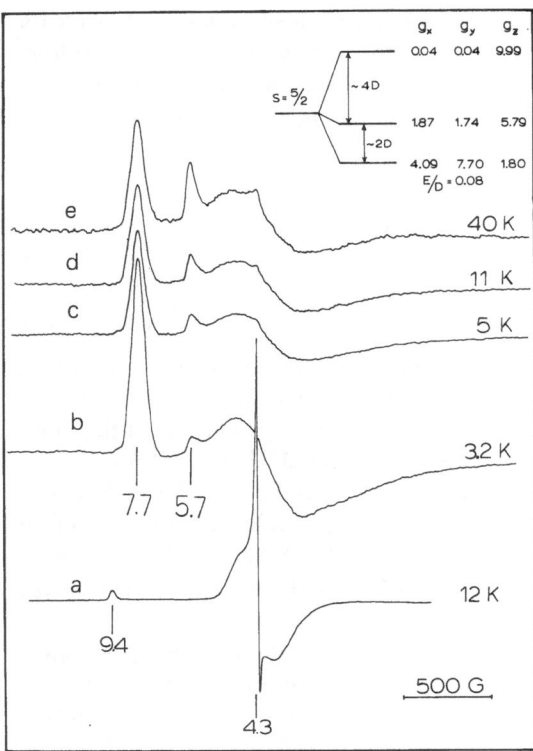

Fig. 1(a). EPR spectrum of *D. gigas* rubredoxin. The spectrum was recorded at the following instrument settings: microwave power, 20 mW; frequency 9.25 GHz; modulation amplitude 10 G; temperature 12 K and gain 100; b–e. EPR spectra of *D. gigas* desulforedoxin at the indicated temperatures and the following powers and gains: (b) and (c) 20 µW, 6.3 × 10³; (d) 20 µW, 10⁴ and (e) 0.2 mW, 2 × 10⁴. The spectra were recorded at the following instrument settings: frequency 9.25 GHz and modulation amplitude 10 G

194

Table 4. Mössbauer parameters of rubredoxin type proteins

	D. gigas desulforedoxin[a]		Cl. pasteurianum rubredoxin[b]	
	Oxid	red	oxid	red
D (cm^{-1})	+ 2.2	− 6	+ 1.92	8.6
ΔE_Q (mm/s)	− 0.75	3.55	− 0.50	− 3.25
δ (mm/s)	0.25	0.70	0.32	0.70
E/D	0.08	0.19	0.23	0.28
H_{sat} (KG)	− 385		− 410	

a Moura et al.[52)]
b Schulz and Debrunner[72)]

at g = 5.7 decreases. The EPR spectra were interpreted[52)] using the spin Hamiltonian for high-spin ferric ion (S = 5/2). The calculated g-values associated with the three Kramers doublets, for the zero field splitting parameter D > O and E/D = 0.08, are shown in Fig. 1. The resonances at 7.7, 4.1 and 1.8 are assigned to the ground state doublet (± 1/2) and the resonance at g = 5.7 to the middle doublet (± 3/2).

The zero field splitting parameters determined for oxidized desulforedoxin (D = 2.2 ± 0.2 cm^{-1} and E/D = 0.08) indicates geometrical differences between this protein and rubredoxin (E/D = 0.23)[72)] (see Table 4). The saturation fields calculated for desulforedoxin (− 385 KG) and rubredoxin (− 410 KG) indicate more covalent bonding in the first protein. In the reduced form the isomeric shift of desulforedoxin is typical of high-spin ferrous ion (S = 2) with tetrahedral sulfur coordination. As shown in Fig. 2, the orbital ground state of desulforedoxin has predominantly a $d_{x^2-y^2}$ character ($\Delta E_Q > O$ and $\eta < 0.5$) while rubredoxin is

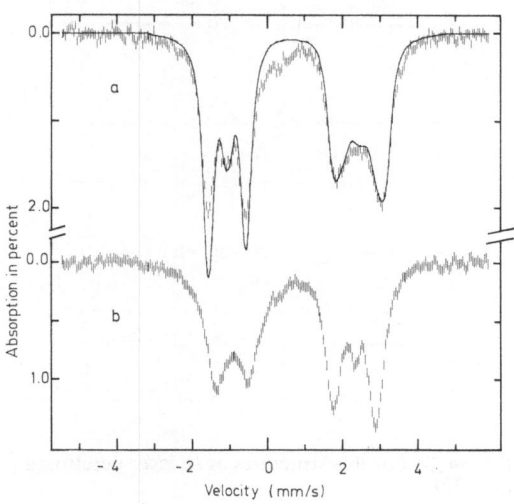

Fig. 2a, b. Mössbauer spectra of reduced *D. gigas* desulforedoxin (a) and *D. gigas* rubredoxin (b) taken at 200 K in a parallel field of 55 KG. Note that the triplet and doublet patterns are reversed for the two proteins[52)]

195

predominantly d_{z^2} ($\Delta E_Q < 0$ and $\eta < 0.5$) causing the inversion of the triplet and doublet pattern. Moura et al.[52] interpreted the differences observed in the electronic structures of the chromophore of desulforedoxin and rubredoxin as the result of different stereochemical constraints imposed by the polypeptide chain.

The comparison of the amino-acid sequence of the two proteins[8,12] can give a clue to the nature of these differences. The spacing of the four cysteine residues in sequences of the type Cys(6)-a-b-Cys(9)-Gly and Cys(39)-c-d-Cys(42)-Gly as observed in rubredoxin are replaced by the sequences Cys(9)-x-y-Cys(12)-Gly and Cys(28)-Cys(29)-Gly. This unusual arrangement may impose the stereochemical constraints which are responsible for the spectral differences between desulforedoxin and rubredoxin. The existence of a dimer structure opens also the question of the possibility that the four cysteines that bind the active center do not belong to the same polypeptide chain, as schematically proposed in Fig. 3[73]. These seem to be the only two possibilities since the two iron atoms should be equivalent according to the spectroscopic data.

In low salt conditions the quantitation of desulforedoxin EPR signals gives approximately 0.5 spin/iron atom. The observations on the spin quantitation of this protein will be discussed here in some detail as they may be of general interest to the EPR study of other proteins. Moura et al.[52] observed that in the oxidized form of desulforedoxin at low salt concentration, in addition to the magnetic Mössbauer spectrum there was a quadrupole doublet with $\Delta E_Q = 0.75$ mm/s (up to 50% of total intensity) which was attributed to fast relaxing material coming from aggregation of the protein. This was concluded by the following reasons: a) the isomeric shifts of the doublet material and magnetic components were the same ($\delta = 0.25$ mm/s); b) samples with different concentrations of doublet material have the same optical spectrum; and c) the Mössbauer spectrum was the same for the doublet and magnetic

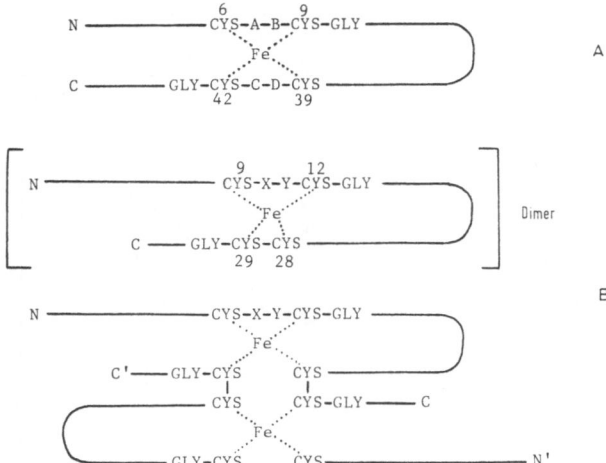

Fig. 3A, B. Possible structures of *D. gigas* desulforedoxin based on analysis of the primary structures[73]

components at $\geqslant 10$ KG. However increasing the ionic strength decreases the quantity of doublet material observed by Mössbauer and increases the number of spin per iron atom quantified by EPR up to approximately 1 spin/iron, as shown in Fig. 4. Thus aggregation can contribute to an underestimation of spin concentration. The aggregated material can be detected both by the presence of fast relaxing Mössbauer signals and the underestimation of the EPR spin quantitation and could not be detected by any other method at room temperature.

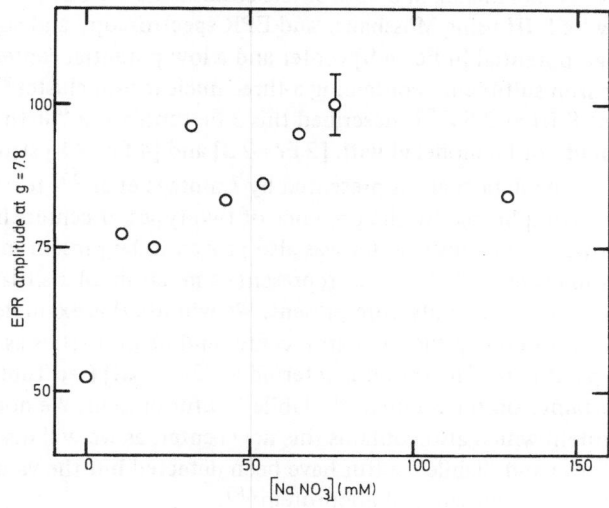

Fig. 4. Dependence of the EPR amplitude of the EPR signal of *D. gigas* desulforedoxin at g = 7.8 on the ionic strength [52]

II. [3 Fe—xS] Centers, a New Basic Structure

The possibility of the existence of new basic structures such as the [3 Fe—xS] core, for the active center of iron-sulfur proteins had a tremendous impact. The new structure was firmly established in the iron-sulfur protein isolated from *A. vinelandii*[27] and in a ferredoxin from *D. gigas*[35]. The *A. vinelandii* iron-sulfur protein designated either as iron-sulfur protein III (AV Fe—S III) by Shetna[70] or as ferredoxin I by Yoch and Arnon[84] (in the text we use the first designation) contains 7—8 iron-atoms and exhibits the EPR and absorption features similar to those of proteins with [4 Fe—S] centers. Sweeney et al.[79] and Yoch and Carithers[86] reported potentiometric titrations of the protein and suggested the presence of two [4 Fe—4 S] centers: one of the centers present in the native form of the protein gives and "isotropic" EPR signal. Another "isotropic" signal, almost superimposable with the first one, appears upon oxidation of the protein with ferricyanide. EPR quantitations give approximately 1 spin/mol for the protein in the native state and 2 spin/mol in the ferricyanide treated protein. The workers proposed that the two clusters were of the "HiPIP" type using the + 3/+ 2 states but with midpoint redox potentials 770 mV apart (+ 350 and − 420 mV were the redox potentials determined). In a further attempt at the identification of the clusters of Av Fe–S III, Howard et al.[32] interpreted the extrusion experiments on the basis of two [4 Fe–4 S] centers, but Averill et al.[5] could not find evidence for more than one [4 Fe–4 S] center. Although EPR strongly suggested the presence of [4 Fe–4 S] centers, the use of complementary techniques proved this assumption to be wrong. Indeed, preliminary X-Ray results at 4.0 A by Stout[76] suggested the existence of two types of centers: one [4 Fe–4 S] and one [2 Fe–2 S]. Meanwhile Münck and co-workers studied the high and low pontential centers of Av Fe-S III using Mössbauer and EPR spectroscopy and suggested the presence of a high potential [4 Fe–4 S] center and a low potential center arranged in a novel type of iron-sulfur core containing a three-nuclear iron cluster[27]. X-Ray studies of Av Fe–S III at 2.5 A[77] described this 3 Fe center as a "distinctly planar structure that could not be modeled with [2 Fe–2 S] and [4 Fe–4 S] structures".

The data analysis presented by Emptage et al.[27] for the Av Fe–S III protein was complicated by the presence of two types of centers in the same molecule. However the novel center was also proven to be present in a ferredoxin isolated from *D.gigas*[35]. This case represents a much simpler situation since the new type of center is the only core present. We will use this example as a prototype of proteins containing the three-iron center and its properties as representative of this new type of core. The center is referred as [3 Fe–xS] (see Table 1 and 3) due to the uncertainty on the number of "labile" sulfur present. We note that in aconitase (a protein which also contains this new center, as we will discuss later) equal amounts of iron and "labile" sulfur have been detected but the values obtained are quite low due to the presence of apoprotein[68].

The ferredoxin isolated from *D.gigas* is present in different oligomeric forms of the same basic unit of M_r = 6.000[11]. Ferredoxin II (FdII) and Ferredoxin I (FdI) are two of the oligomeric forms in which the ferredoxin of this organism can be isolated[11]. In particular FdII is a tetramer (M_r = 24.000) and FdI is a trimer (M_r = 18.000). Although the primary sequence of the basic unit of both oligomers has been verified to be identical[10] the two proteins differ in optical properties, physiological activities and oxidation-reduction properties[11,17,49].

As isolated FdII exhibits a fairly isotropic EPR (Fig. 5) signal around g = 2[17] and the temperature dependence studies suggest that the material is homogeneous and only one EPR active species is present[35]. A simulation of the spectrum was obtained with g_1 = 2.02, g_2 = 2.00 and g_3 = 1.97, using gaussian widths of 15,35 and 80 G respectively. This EPR spectrum is similar to those observed in other bacterial ferredoxins in the oxidized state[62] and has also been attributed in those cases to the + 3 state of the [4 Fe–4 S] center[78]. *D.gigas* FdII is an example of how EPR techniques can be misleading in the identification of new iron-sulfur centers.

Iron determinations on FdII give systematically approx. 3 iron atoms per monomeric unit and the quantitation of the EPR signal accounts for 0.91–0.95 spin/3 Fe atoms[35]. In the fully reduced form the features of the g = 1.94 EPR type signal are very weak in FdII (0.02–0.03 spins/3 Fe atom) and attempts to further reduction were not successful[17].

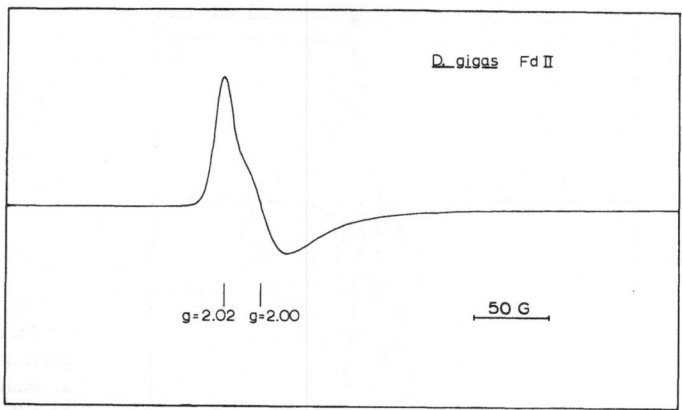

Fig. 5. EPR spectrum of oxidized (native) *D. gigas* ferredoxin II, 0.180 mM in monomer. Spectrum recorded at the following instrument settings: microwave power 30 µW; frequency 9.218 GHz; modulation amplitude 4 G; temperature 8 K and gain 1.250

The oxidized form of FdII (Fig. 6) shows a magnetic Mössbauer spectrum at 4.2 K with three distinct iron sites. The spectrum components 1 and 2, were also observed in the Av Fe–S III protein [27] but the component 3 could not be easily depicted since the absorption of the high potential center masks this spectral region. The three components belong to the EPR active center [35], in agreement with EPR quantitation and the presence of a unique EPR active species. Indeed, high temperature studies of the oxidized material (native material) at 77 K reveals that the three

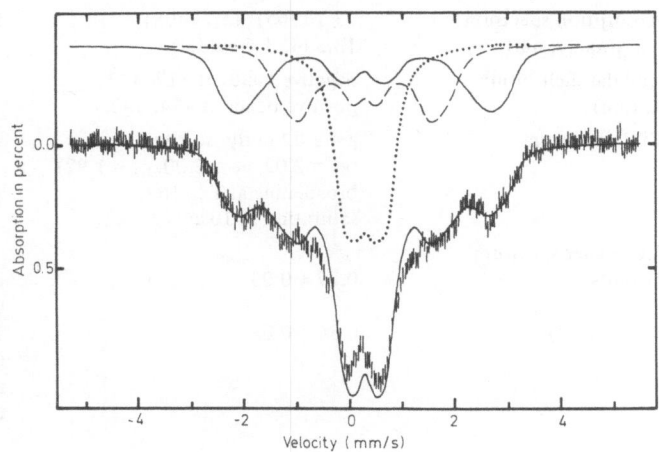

Fig. 6. Mössbauer spectrum of oxidized *D. gigas* ferredoxin II taken at 1.5 K in a field of 600 G applied parallel to the observed γ-radiation. The solid line plotted over the data is a superimposition of three simulated spectra describing components 1, 2 and 3 [35]

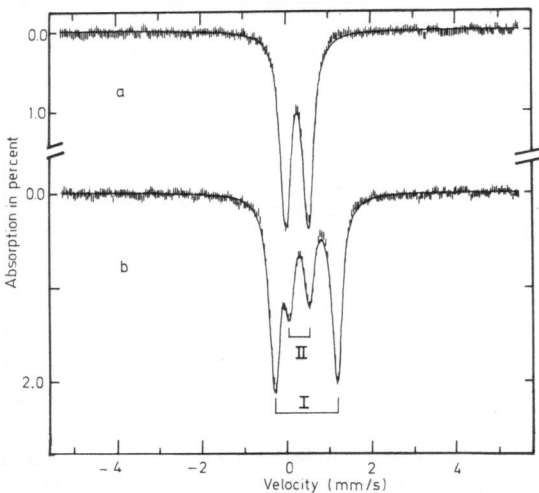

Fig. 7a, b. Zero-field Mössbauer spectrum of *D. gigas* ferredoxin II (a) oxidized form at 77 K (b) reduced form at 4.2 K.[35] Spectrum (b) will be referred as the "finger print" spectrum of the new iron center [3 Fe−xS]

components give the same Mössbauer spectrum (Fig. 7) with isomeric shifts and quadrupole doublet values which are typical of tetrahedral coordination by sulfur ligands (see Table 5). Upon one electron reduction an EPR silent state is observed and two distinct environments could be detected by Mössbauer spectroscopy at zero field (Fig. 7), designated as doublet I and doublet II with a ratio of 2:1 (see

Table 5. Characteristics of the novel [3 Fe−xS] center (*D. gigas* FdII is considered the prototype)

Property	Oxidized form	Reduced form
Absorption spectrum $\epsilon_{mm}(\lambda_{max}, nm)$	23.1 (305) 15.7 (405) 13.3 (453 shoulder)	9.6 (425 shoulder)
Circular dichroism[a] λ, (nm)	negative bands at 317, 423 positive bands at 474, 580	
EPR spectrum	g = 2.02 fairly isotropic (g_z = 2.02, g_y = 2.00, g_z = 1.97) broadening at T > 16 K saturation at $10 \mu w$ (2.7 K)	no signal[b]
Mössbauer spectrum		
δ (mm/s)	0.27 ± 0.03	0.46 ± 0.02 (doublet I)[c] 0.30 ± 0.02 (doublet II)
ΔE_Q (mm/s)	0.54 ± 0.03	1.47 ± 0.03 (doublet I) 0.47 ± 0.02 (doublet II) ratio doublets I/II ≅ 2.00 Characteristic broadening at field > 600 G

[a] In general features the CD spectrum of FdII does not resemble [4 Fe−4 S] centers except in their intensity[17]

[b] A very weak 1.94 type signal can be observed in FdII (less than 2%)[17]

[c] Doublet I and II are shown in Fig. 7

also Table 5). The doublet II component is high spin ferric in character (see Möss-bauer parameters in Table 5) and the application of magnetic fields to the reduced material lead Huynh et al. [35] to conclude that it must belong to a spin coupled system. Furthermore, doublet I contains two equivalent iron sites that remain indiscernible in strong magnetic fields (10 KG).

Putting together these data it was concluded that: a) in the oxidized state the three iron atoms are magnetically distinct and b) in the reduced form two iron sites are indiscernible and the third iron site (doublet II) is high-spin ferric in character. These conclusions suggest that the electron picked up by reduction is shared between the two iron atoms responsible for doublet I. The increase in δ and ΔE_Q of the iron atoms of doublet I upon reduction, points to a formal oxidation state of + 2.5 for these two iron atoms (see also Table 7).

An interesting fact is that when FdII is converted into the apoprotein by treatment with trichloroacetic acid and the iron-sulfur center is reconstituted by the addition of iron and sulfide [17] the protein becomes a [4 Fe—4 S] ferredoxin. Indeed the isotropic type signal of the oxidized form is no longer observed and upon reduction with dithionite a g = 1.94 type signal indicative of a + 1 oxidation state is observed at 4.2 K (see Fig. 8). This reconstituted material exhibits Mössbauer spectra almost identical to the observed for the reduced ferredoxin from *Bacillus stearother-mophilus* [46] a protein for which a [4 Fe—4 S] center is firmly established (Fig. 9).

The oligomeric form FdI exhibits, as shown in Fig. 10, an "isotropic" type signal in the oxidized form which may account for 10—25% of the total intensity (the size of the "isotropic" signal varies with different preparations). Upon reduction it exhibits a strong EPR signal with principal g values at $g_1 = 1.92$, $g_2 = 1.94$ and

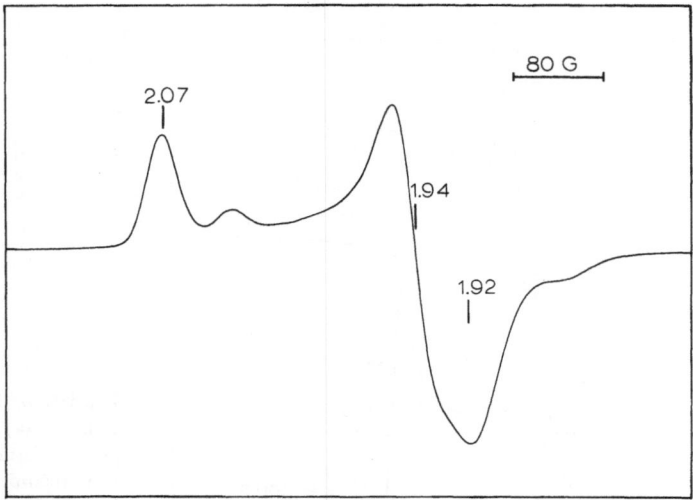

Fig. 8. EPR spectrum of reconstituted *D. gigas* ferredoxin II in the reduced form. The spectrum was recorded at the following instrumental settings: microwave power 2 mW; frequency 9.223 GHz; modulation amplitude 10 G; temperature 18 K and gain 1000

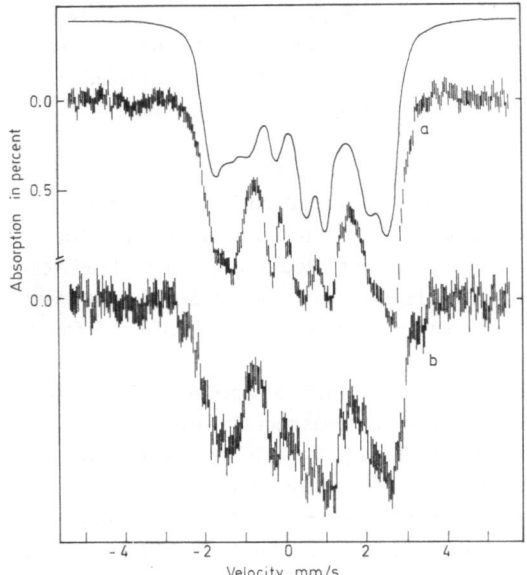

Fig. 9.a Mössbauer spectrum of reduced reconstituted *D. gigas* ferredoxin II taken at 4.2 K in a field of 600 G applied parallel to the observed γ-radiation; b Mössbauer spectra of reduced *D. gigas* ferredoxin I (subtraction of 25% of a spectrum of a three-iron center run in the same conditions was made). Spectrum obtained in natural abundance Fe[56]. Experimental conditions as for spectrum (a).
The uper trace is the simulated Mössbauer spectra obtained using *B. stearothermophilus* [4 Fe−4 S] ferredoxin parameters for the reduced form[46] and ΔE_Q and δ values determined for reduced reconstituted *D. gigas* ferredoxin II. (Our unpublished results in collaboration with Drs. B. H. Huynh and E. Münck)

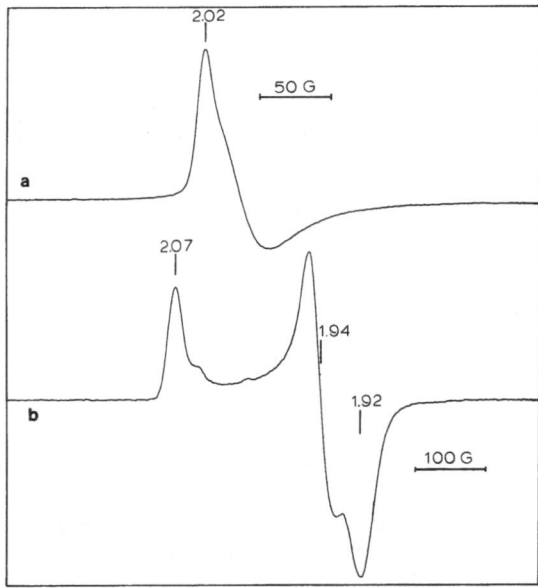

Fig. 10a, b. EPR spectra of oxidized (a) and reduced (b) *D. gigas* ferredoxin I. The spectra were recorded at the following instrumental settings: microwave power 10 µW; frequency 9.220 GHz; modulation amplitude 4 G; temperature 8 K and gain 4000 (a) and 8000 (b)

$g_3 = 2.07$ [17] (Fig. 10). The intensity of the EPR signal in the reduced form accounts for the missing intensity of the oxidized form. At 4.2 K this form has a magnetic Mössbauer spectrum almost superimposable with the Mössbauer spectrum obtained for the reconstituted FdII in the reduced form (see Fig. 9). Upon oxidation the $g = 1.94$ disappears and the Mössbauer spectrum changes to give a quadrupole pattern. High field studies prove that this pattern results from iron atoms in diamagnetic sites confirming the presence of [4 Fe–4 S] centers as the major component of FdI (our unpublished data).

The results obtained with the oligomeric forms of *D.gigas* ferredoxin and reconstituted material, implicate that the same amino-acid polypeptide chain can accomodate both [3 Fe–xS] and [4 Fe–4 S] centers. FdII represents a very pure and homogeneous preparation in respect to the type of centers present, being an example of a protein containing only [3 Fe–xS] cores. The oligomeric FdI is isolated in a form that contains both molecules with one [4 Fe–4 S] center and a small number (up to 25%) of molecules with one [3 Fe–xS] center. This situation is different from that of the AvFe–S III protein where in each molecule one center of the novel structure is associated with one [4 Fe–4 S] center of "HiPIP type".

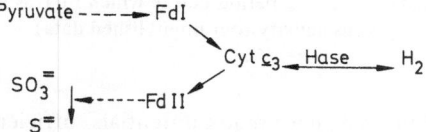

Fig. 11. The role of the oligomeric form of *D. gigas* ferredoxin in metabolic pathways of sulfate reducing bacteria

D.gigas FdII and FdI have been implicated in two important methabolic pathways of this sulfate reducing organism [48,49], the phosphoroclastic reaction [6] and the sulfite reductase system [39] schematically shown in Fig. 11. The structural studies described above should be analysed in perspective to the redox potentials and physiological activities of the two oligomeric forms:

i)   For FdII the redox potential value of the [3 Fe–xS] center is – 130 mV. For FdI the redox potential value of the [3 Fe–xS] center is – 50 mV and that of [4 Fe–4 S] is – 455 mV [17].
ii) FdII is more efficient than FdI in the sulfite reductase electron transport system [11].
iii) FdI is active as an electron carrier in the phosphoroclastic reaction [49]. In the same conditions FdII is not active. However, after a long time lag phase FdII also stimulates this reaction. These results are summarized in Fig. 12 in the form of concentration saturation curves.

Put together, these observations give some clues in the determination of a specific task for the novel structure.

As the redox potential of sulfite reductase is about – 110 mV and that of pyruvate dehydrogenase is about – 600 mV it is reasonable that the [3 Fe–xS] centers, which have less negative redox potentials are the most efficient in ii) and that only the

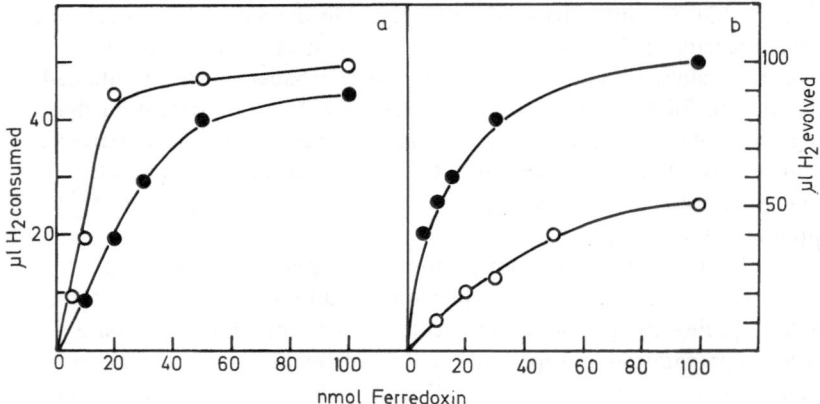

Fig. 12a, b. Stimulation of electron transport in two metabolic pathways of *D. gigas* by ferredoxin. (a) Sulfite reductase activity dependence on Fd II (o) and Fd I (•) concentration. The values of hydrogen comsumption represent the activity after 12 min and were corrected for the endogenous activity (adapted from Bruschi et al.,[8,9]). (b) Pyruvate dehydrogenase activity dependence of Fd II and Fd I concentration. The values determined for hydrogen evolution represent the activity after 12 min (between 8−20 min for Fd I and between 28−40 min for Fd II, due to the existence of a lag period before which Fd II is not active[49]. The values were corrected for the endogenous activity (our unpublished data)

[4 Fe−4 S] have redox potentials sufficiently low to be active in the phosphoroclastic reaction. This would implicate that the [3 Fe−xS] centers of FdII can be interconverted into [4 Fe−4 S]. This interconversion is supported by the fact that FdII is only active in the last system after a lag time phase and that [4 Fe−4 S] are obtained when reconstituting FdII.

The fact that the same amino acid polypeptide chain can accomodate these two types of cores raises several questions: what is the process that regulates the building up of a three or four iron core? Is there interconversion between the two structures (as suggested by reconstitution experiments and participation of FdII in the phosphoroclastic reaction)? Is there a specific biological role for each type of structure? Are [4 Fe−4 S] structures easier to build up? But if so, why are [3 Fe−xS] such stable structures (generally these structures are thermostable and resistant to chemical denaturation)? The answer to these exciting questions needs further basic biochemical work for which the *Desulfovibrio gigas* methabolic pathway scheme is a suitable system.

The novel [3 Fe−xS] core with three covalently linked iron atoms seems to be a wide spread structure since it can be found in proteins as different as beef hearth aconitase and glutamate synthase (Huynh and Münck, personal communication). All these proteins give a typical Mössbauer spectrum in the reduced form identical to that obtained for *D.gigas* FdII. Aconitase contains only one type of iron-sulfur center. While ferredoxins are only involved in electron transfer, aconitase has a central metabolic role in the citric acid cycle[37,68]. The role of the iron-sulfur center remains to be determined. Glutamate synthase is an important enzyme in the nitrogen metabolism. However it contains several prosthetic groups in addition to the three iron center [65] and again the role of this center is not known. So far of all proteins containing only Fe—S centers only for the case of hydrogenase has a catalytic activity of the center been discussed.

A survey of the literature suggests the presence of this novel structure in other proteins. Likely candidates, judged by the features of EPR spectra and other general properties, such as the redox potential are: *Mycobacterium flavum* FdI [83]; *Thermus thermophylus* Fd [57]; *Methanosarcina barkeri* Fd (our unpublished data); Mitochondrial succinate dehydrogenase [56]; *Rodospirillum rubrum* Fd IV [87]; and *Desulfovibrio gigas* hydrogenase (our unpublished data in collaboration with J. Le Gall).

Further characterization of these proteins in the light of the work reported here, as well as progress in the search of synthetic analogues of this novel cluster, are awaited with great interest.

III. Nitrogenase Centers, Complex Centers for a Specific Job

The understanding of a biochemical process as fundamental as nitrogen fixation attracted several research groups to study the nitrogenase enzyme from different points of view. A combination of Mössbauer and EPR techniques has been used to reveal the redox processes and the oxidation states involved [33,34,54,74,88]. The Mo—Fe protein contains approximately 30 iron atoms and 2 molybdenum atoms as judged by comparing preparations from different sources [61,69].

The iron atoms are associated in two groups of centers which present novel features when compared with well established iron-sulfur centers (see Table 6):
a) two M-centers which contain 1 molybdenum and 6 iron atoms each. They are magnetic (M) in the native form and are identical in spectroscopic terms to the FeMo cofactor isolated by Shah and Brill [69].
b) four P-clusters containing 4-iron atoms each, which may represent a variation of the basic [4 Fe—4 S] structure, where a different ligation mode or new types of ligands around the core have been suggested [61].

The cofactor and the P-cluster are distinct entities as shown by the fact that in the presence of thiophenol and 80% N-methyl formamide solutions, *A. vinelandii* gives rise to cofactor and $(\phi S)_4 F_4 S_4$ signals [61].

1. M-Clusters

The Mo—Fe protein in the native state exhibits an EPR spectrum due to two S = 3/2 centers [54] which were shown to be identical to the signals observed for the Fe—Mo cofactor [64] with features around g = 4.3, 3.1 and 2.0 for the native enzyme. These M-centres contain about 40% of the total iron [54,74] which were associated with the S = 3/2 centers, as judged by conjunction of EPR and Mössbauer methods. The Mössbauer spectrum of both the S = 3/2 species and the cofactor, were studied in detail by Huynh *et al.* [33] and the spectral features could be decomposed in 6 subcomponents suggesting the existance of 12 iron atoms in 2 cofactors (S = 3/2)

Table 6. Characteristics of the iron cluster arrangement in nitrogenase [a]

Components seen by Mössbauer	Number of iron atoms associated to each component based in a total of approx. 30 iron atoms	
M	12 ± 2	
P { D		$12-13$
16 ± 1		
Fe^{2+}		4
S [b]	2	

Oxidation states and spin electronic states at M and P clusters [c]
Co-factor cluster M

$$M^{ox} \underset{1e}{\rightleftarrows} M^N \underset{1e}{\rightleftarrows} M^R$$

$S = 0 \qquad S = 3/2 \qquad S = \text{integer}$

P-clusters

$$P^{ox} \underset{1e}{\rightleftarrows} P^N$$

$S \geqslant 3/2 \qquad S = 0$

[a] Table adapted from Münck et al.[54], Zimmerman et al.[88]
[b] S component revealed to be inert in all protein preparations studied[88]
[c] ox, N and R state for oxidized, native (as isolated in the presence of dithionite) and reduced enzyme

structures. The M-clusters can be stabilized in three oxidation states as indicated in Table 6[88] and the 6 iron atoms were shown to belong to a spin coupled structure[33]. In the native form (M^N) the electronic ground state has a spin $S = 3/2$ and oxidation produces a diamagnetic state, $S = O$. Under nitrogen fixing conditions the center M is in the reduced form[55].

The chemical nature of the M-clusters is not known. It seems to involve 1 Mo, 6 iron atoms, and labile sulfur. The presence of any associated organic material has not yet been firmly established. These centers seem to be arranged in a novel structure and it is suggested that they have direct implication in the catalytic active site of the enzyme[61]. This unique structure has only been found in nitrogen fixing processes and represents a core performing a very specific task. Model compounds that integrate Mo, Fe and S atoms in unique structures have been extensively studied[23,75] but the way they mimic the nitrogenase center is not yet clear.

2. P-Clusters

Evidence for the presence of another type of centers was obtained by oxidation of the enzyme with thionine[88] which reveals that the oxidation takes place in two distinct steps: firstly the protein can be oxidized, taking four electrons without loss of the characteristic EPR spectrum of the cofactor, $S = 3/2$ (M^N); and secondly a two electron process takes place leading to the disappearance of the EPR signal ($M^N \rightarrow M^{ox}$). The data suggest that the two components D and Fe^{2+} following the nomenclature of Münck et al.[54], which contain approximately 50% of the total iron (16 iron atoms) are the centers undergoing the four electron oxidation since the Mössbauer spectrum of these components is converted, during the oxidation, from quadrupole doublets into a magnetic spectrum. Again these two components, D and Fe^{2+}, might be associated in a structure of a novel type designated as P-clusters[54,59]. Mössbauer studies suggest that the D centers contain low-spin ferrous ions. The D irons are in a diamagnetic complex, but participating in spin coupled clusters with the Fe^{2+} component. The iron of the Fe^{2+} component is high-spin ferrous in character and is coordinated by sulfur. They participate in a spin coupled cluster[33] and by oxidation it gives an $S \geqslant 3/2$ species. The intensity ratio of $3:1$ observed between the D and the Fe^{2+} groups suggests the presence of $4 \times [4\,Fe-4\,S]$ units (P-clusters) but these centers are somewhat distinct from typical $[4\,Fe-4\,S]$ centers. The P-clusters are intriguing structures since the isomeric shift observed in the P^N states is 0.64 mm/s[33]. The characterization of these oxidation states according to the Carter et al.[21] hypothesis should be as the $+0$ state[44,61] since reduced ferredoxin has a $\delta = 0.50-0.58$ mm/s and the change between $+3/+2$ and $+2/+1$ states introduces an increase of 0.1 mm/s in the observed isomeric shift (see Table 7). The presence of such a state has not yet been demonstrated in biological systems.

P-clusters were observed in two different oxidation states: they are isolated in a diamagnetic state (P^N) and can be converted into a paramagnetic form (P^{ox}) upon one electron oxidation.

Orme-Johnson and Münck (1980) in unpublished data refer to transients signals with $g \cong 2$ (associated with a redox potential near -240 mV) which are observed during the oxidation of these clusters, supporting the existence of P^N in the most reduced state. However, the problem may be more complex since ferredoxin type ($g < 2$) and HiPIP type ($g > 2$) EPR signals are observed in CO inhibited conditions. Another possibility could be that the transition is $+3/+2$ but the redox potential does not seem appropriate.

Table 7. Selected Mössbauer isomeric shifts of iron sulfur proteins[a]

Protein	δ (mm/s)	Oxidation state
Fe^{3+} rubredoxin	~ 0.25	
oxid *Chromatium* HiPIP	0.31	+3
red *Chromatium* HiPIP		
or	0.42	+2
oxid *B. stearothermophilus* Fd		
red *B. stearothermophilus* Fd	0.50–0.58[b]	+1
Fe^{2+} rubredoxin	0.65	
P-cluster nitrogenase	0.64	(D component)
	0.64	(Fe^{2+} component)

[a] Table adapted from Cammack et al.[18] and Huynh et al.[34]
[b] Two components are well resolved in the Mössbauer spectra[18]

C. How to Identify Fe–S Centers

The fully characterization of both simple or more complex (but well defined in terms of active center composition) iron-sulfur proteins leads to a well of information. The compilation of typical spectroscopic features of the known basic iron-sulfur structures enables a preliminary characterization of centers in a new simple situation or even in some more complex ones. EPR spectroscopy of the iron-sulfur cores in the appropriated oxidation states have characteristics that can be used to readily distinguish certain type of centers[58]. This technique has also been used to analyse components in complex systems. However the use of EPR as the sole technique can be misleading, when applied to new situations as we have seen for the case of the [3 Fe–xS] core. In this particular case only the conjunction of EPR and Mössbauer can lead to a proper characterization[27,33,34,55].

Extrusion and "core transfer" techniques can give information when applied to [2 Fe–S] and [4 Fe–S] systems, but should be applied with extreme care so that no interconversion occurs (see detailed discussion in Ref. 30 and 60).

Other techniques, like U.V./visible, CD and MCD spectroscopy can be useful when applied to simple systems specially to rubredoxin and [2 Fe–2 S] type protein[29].

Although elementary analysis of iron and labile sulfur may appear at first sight to be a linear problem, the determination of reliable values is extremely difficult due to several facts including the protection of the active center by the apoprotein, adventitious metal binding sites and instability of the metal core. The knowledge of the protein concentration and the molecular weight is indispensable for correlating the quantitative analysis of the metal of the active center. For small proteins these parameters can be well defined, specially when the protein concentration is related to the amino-acid analysis of stable amino-acid residues, but is not always easy for larger molecules.

Only the conjunction of different techniques can give a full characterization of the type of active center:

[Rb] *center:* U.V./visible and CD spectroscopy show characteristic bands. Reduction with ascorbic acid can usually be accomplished. EPR has characteristic features at $g = 4.3$ readily observable at 77 K with additional signals at $g = 9.4$ at $T < 20$ K, in the oxidized form. No signal is observed in the reduced state.

[2 Fe–2 S] *center:* Characteristic U.V./visible and very intense CD spectra. EPR variable from narrow (putidaredoxin) to broad and rhombic (plant ferredoxin), $g < 2$ signal in the reduced state, observable at 77 K. No signal in the oxidized state. Isotropic EPR signal observed below 77 K in 80% DMSO in reducing conditions. Extrusion and displacement reaction give characteristic products.

[3 Fe–xS] *center:* U.V./visible typical, similar to [4 Fe–4 S] centers. CD different from [2 Fe–2 S] and [4 Fe–4 S] but not readily identifiable. EPR in oxidized state, fairly isotropic centered around $g = 2.02$. No signal in the reduced state. Wide range of redox potentials. Typical Mössbauer spectra in the reduced form at 4.2 K in zero field (see Fig. 7B) with two doublets in the ratio 2:1. A substantial broadening at weak applied fields (600 G) might be a typical feature. The extrusion techniques so far performed give misleading results. Unenriched material containing as less as 2 μmoles of iron in a 0.3 ml sample is sufficient in order to obtain the Mössbauer spectrum that can be used as a fingerprint of the [3 Fe—xS] core.

[4 Fe—4 S] *center:* U.V./visible with broad bands around 400 and 300 mm, but similar to those observed for [3 Fe—xS] centers. Weak CD spectrum not easily identifiable. Extrusion and displacement reactions give characteristic products. EPR in 80% DMSO (reducing conditions) give an axial type spectrum observable below 35 K. HiPIP and ferredoxin type centers may be readily distinguished based on EPR characteristics. So far all the HiPIP known have a very positive redox potential.

D. Conclusions. Is There a Correlation Between Redox Potential and the Type of Fe—S Center?

The values of midpoint oxidation-reduction potential are essential parameters for the characterization of an iron-sulfur protein as well as for the determination of their physiological function. However the correlation of redox potential with the type of active center is not straighforward.

Rubredoxin type proteins constitute the only group with a narrow range of redox potentials centered around 0 mV. A value of -57 mV was determined by Lovenberg and Sobel[43] for *Cl.pasteurianum* rubredoxin. The redox potential values of rubredoxins isolated from sulfate reducing organisms[51] (*D.gigas*, Rb, $+6$ mV, *D.salexigens* Rb, -31 mV, and *Drm.acetoxidans*, -46 mV) and two rubredoxins isolated from *Cl.thermoaceticum*[82] were recently reported. Spectroscopic studies have shown a close similarity for the redox centers of rubredoxins (see for example the comparison between Mössbauer data of *D.gigas*[52] and those of *Cl.pasteurianum* rubredoxin[25]. The primary sequence studies available for rubredoxins isolated from *D.gigas*[9] and *Cl.pasteurianum*[45], as well as the X-Ray structures obtained for *Cl.pasteurianum* Rb[80] and *D.vulgaris* Rb[3] indicate that the main dissimilarities are the number and position of charged residues which may modulate structural features of the active center in a very subtle way. The importance of the differences in charged residues may also explain the high specificity of NADH—H$^+$ rubredoxin-oxidoreductase activity of *D.gigas* for the rubredoxin of the same organism ($K_M = 6.2 \times 10^{-6}$M) when compared with rubredoxin of other similar organisms, *D.vulgaris* Rb ($K_M = 5.3 \times 10^{-5}$M) and *Cl. pasteurianum* Rb ($K_M = 1.0 \times 10^{-4}$M) (Odom and Peck unpublished results Ref. 40).

For the other groups of proteins the situation is more complex. In the case of the [4 Fe—4 S] centers the "three state" hypothesis[21] correlates the oxidation state with the redox potential (see Table 2). However, the validity of this hypothesis was contested by reports which wrongly assigned redox transitions $+3/+2$ to the negative redox potential determined for one of the centers in *A.vinelandii* Fe—S III[79] and for *D.gigas* FdII[17]. Later characterization of these centers as [3 Fe—xS] centers together with the available data for the other centers allows the following conclusions regarding redox potentials of iron-sulfur proteins (see Table 8):

Table 8. Range of redox potentials in iron-sulfur proteins

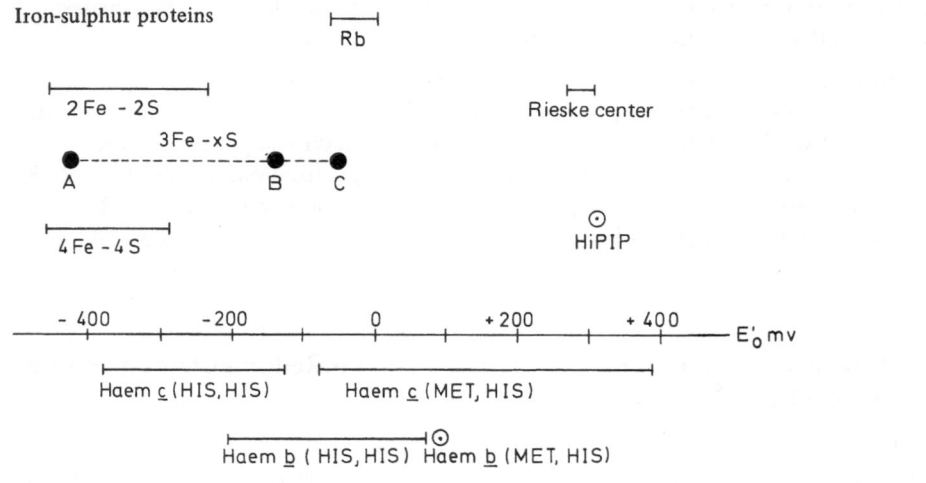

Haem proteins

Table adapted from Rieske[66], Lovenberg and Sobel[43], Moura et al.[51], Cammack[14], Cammack et al.[16,18] and Yoch and Carithers[85]

The values shown for [3 Fe—xS] centers are:

(A) AvFe-S III [79]

(B)(C) *D. gigas* FdII and I respectively (redox potential associated with the "isotropic" type signal)[17]

1) Rubredoxin type proteins have a narrow range of redox potentials around 0 mV. The value determined for *D.gigas* desulforedoxin is within this range (-35 mV)[50].
2) [2 Fe—2 S] centers cover a wide range of negative redox potentials (-450 to -220 mV) excluding the Rieske centers[66].
3) The redox transition $+2/+1$ in [4 Fe—4 S] proteins has always been associated with a negative redox potential value.
4) The range of redox potentials observed for [4 Fe—4 S]$^{+2(+2,+1)}$ centers overlap with the redox potentials observed for [2 Fe—2 S] centers.
5) The redox transition $+3/+2$ in [4 Fe—4 S] proteins seems to be associated with a positive redox potential.
6) The [3 Fe—xS] cores present a very wide range of redox potential values. The two cases so far characterized Av Fe—S III and *D.gigas* FdII have both negative redox potentials within a range of approximately 300 mV, but this range may be enlarged when more data is available.
7) The influence of the polypeptide chain on the same active center is notorious, when observing the range of redox potentials available for [2 Fe—2 S] and [4 Fe—4 S] proteins. However, the modulation of the redox potential by the protein structures is a common feature to other groups of proteins. Haem proteins are a good example of this fact since the "rigid" and well-defined structure of the

active center can transfer electrons in a very wide range of redox potentials (see also Table 8).

Acknowledgements. We thank our collaborators, Drs. R. Cammack, B. H. Huynh, J. Le Gall and E. Münck, for fruitful discussions and experimental contributions. This work was supported by NIH Grant no. GM 25879 and the INIC, JNICT and the Calouste Gulbenkian Foundation of Portugal.

References

1. Adman, E. T.: Biochim. Biophys. Acta *549*, 107 (1979)
2. Adman, E. T., Sieker, L. C., Jensen, L. H.: J. Biol. Chem. *248*, 3987 (1973)
3. Adman, E. T., et al.: J. Mol. Biol. *112*, 113 (1977)
4. Averill, B. A., Orme-Johnson, W. H.: Iron-Sulfur Proteins and their Synthetic Analogs. In: Metal Ions in Biology, Vol. 7 (ed. Siegel, H.), New York, Marcel Dekker 1978, pp. 127–183
5. Averill, B. A., Bale, J. R., Orme-Johnson, W. H.: J. Am. Chem. Soc. *100*, 3034 (1978)
6. Akagi, J. M.: J. Biol. Chem. *242*, 2478 (1967)
7. Blumberg, W. E.: EPR of high Spin Fe^{3+} in Rhombic Fields. In: Magnetic Resonance in Biological Systems (eds. Ehrenberg, A., Malmstrom, B. C., Vanngard, T.), Oxford, Pergamon Press 1967, pp. 119–133
8. Bruschi, M.: Biochem. Biophys. Res. Commun. *70*, 615 (1976)
9. Bruschi, M.: Biochim. Biophys. Acta *434*, 4 (1976)
10. Bruschi, M.: Biochem. Biophys. Res. Commun. *91*, 623 (1979)
11. Bruschi, M., et al.: Biochim. Biophys. Acta *449*, 275 (1976)
12. Bruschi, M., et al.: Biochem. Biophys. Res. Commun. *90*, 596 (1979)
13. Cammack, R.: Biochem. Biophys. Res. Commun. *54*, 548 (1973)
14. Cammack, R.: Functional Aspects of Iron-Sulfur Proteins. In: Metalloproteins: Structure, Function and Clinical Aspects (ed. Weser, V.), Stuttgart, Thieme 1979, pp. 162–184
15. Cammack, R., Evans, M. C. W.: Biochem. Biophys. Res. Commun. *67*, 544 (1975)
16. Cammack, R., et al.: Biochem. J. *168*, 205 (1977)
17. Cammack, R., et al.: Biochim. Biophys. Acta *490*, 311 (1977)
18. Cammack, R., Dickson, D. P. E., Johnson, C. E.: Evidence from Mössbauer Spectroscopy and Magnetic Resonance on the Active Centers of the Iron-Sulfur Proteins. In: Iron-Sulfur Proteins (ed. Lovenberg, W.), Vol. 3, New York, Academic Press 1977, pp. 283–330
19. Carter, C. W.: X-Ray Analysis of the High Potential Iron-Sulfur Proteins and Ferredoxins. In: Iron-Sulfur Proteins (ed. Lovenberg, W.), Vol. 3, New York, Academic Press 1977, pp. 157–204
20. Carter, C. W., Jr., et al.: Cold Spring Harb. Symp. Quant. Biol. *36*, 381 (1971)
21. Carter, C. W., Jr., et al.: Proc. Natl. Acad. Sci., U.S.A. *69*, 3526 (1972)
22. Coucouvanis, D., et al.: J. Am. Chem. Soc. *98*, 5721 (1976)
23. Cramer, S. P., et al.: J. Am. Chem. Soc. *100*, 3814 (1978)
24. Cushman, D. W., Tsai, R. L., Gunsalus, I. C.: Biochem. Biophys. Res. Commun. *26*, 577 (1967)
25. Debrunner, P. G., et al.: Recent Mössbauer Results of some Iron-Sulfur Proteins and Model Complexes. In: Iron-Sulfur Proteins (ed. Lovenberg, W.), Vol. 3, New York, Academic Press 1977, pp. 381–417
26. DerVartanian, D. V., et al.: Biochem. Biophys. Res. Commun. *26*, 569 (1967)
26a. Dunham, W. R., et al.: Two-Iron Ferredoxin in Spinach, Parsley, Pig adrenal cortex, *Azotobacter vinelandii*, and *Clostridium pasteurianum*: Studies by Magnetic Field Mössbauer Spectroscopy, Biochim. Biophys. Acta *253*, 134 (1971)

211

A. V. Xavier et al.

27. Emptage, M., et al.: J. of Biol. Chem. *255*, 1793 (1980)
27a. Gibson, J. R., et al.: The Iron Complex in Spinach Ferredoxin, Proc. Natl. Acad. Sci. U.S.A. *56*, 987 (1966)
28. Greenwood, C., Barber, D.: Inorg. Chem. *1*, 210 (1979)
29. Hall, D. O., Cammack, R., Rao, K. K.: Non-Heme Iron Proteins. In: Iron in Biochemistry and Medicine (eds. Jacobs, A., Worwood, M.), New York, Academic Press 1974, pp. 279–334
30. Holm, R. H.: Identification of Active Sites in Iron-Sulfur Proteins. In: Biological Aspects of Inorganic Chemistry (eds. Addison, A. W., et al.), New York, Willey Interscience Publications 1977, pp. 77–111
31. Holm, R. H., Ibers, J. A.: Synthetic Analogues of the Active Sites of Iron-Sulfur Proteins. In: Iron-Sulfur Proteins (ed. Lovenberg, W.), Vol. 3, New York, Academic Press pp. 205–281
32. Howard, J. B., Lorsbach, T., Que, L.: Biochem. Biophys. Res. Commun. *70*, 582 (1976)
33. Huynh, B. H., Münck, E., Orme-Johnson, W. H.: Biochim. Biophys. Acta *576*, 192 (1979)
34. Huynh, B. H., et al.: Nitrogenase XII: Mössbauer Studies of the MoFe Protein from *Clostridium pasteurianum* WS, Biochim. Biophys. Acta in press (1980)
35. Huynh, B. H., et al.: Evidence for a Three Iron Center in a Ferredoxin from *Desulfovibrio gigas*, J. of Biol. Chem. in press (1980)
36. IUB-IUPAC: Nomenclature of Iron-Sulfur Proteins – Recommendations, 1978. Biochim. Biophys. Acta *549*, 101 (1979)
37. Kurtz, D. M., et al.: J. Biol. Chem. *254*, 4967 (1979)
38. Lane, R. W., et al.: Proc. Natl. Acad. Sci., U.S.A. *72*, 2868 (1975)
39. Le Gall, J., Dragoni, N.: Biochem. Biophys. Res. Commun. *23*, 145 (1966)
40. Le Gall, J., DerVartanian, D. V., Peck, H. D., Jr.: In: Flavoproteins, Iron-Proteins, and Hemo-proteins as Electron Transfer Components of the Sulfate-Reducing Bacteria (ed. Sanadi, R.), New York, Academic Press 1979, Current Topics in Bioenergetics *9*, 237
41. Lode, E. T., Coon, N. Y.: J. Biol. Chem. *246*, 791 (1971)
42. Lovenberg, W. (ed.): Iron-Sulfur Proteins, Vol. 1 (1973), Vol. 2 (1973) and Vol. 3 (1977), Academic Press Inc., New York
43. Lovenberg, W., Sobel, B. E.: Proc. Natl. Acad. Sci. U.S.A. *54*, 193 (1965)
44. Lowe, D. J., Eady, R. R., Thorneley, R. N. F.: Biochem. J. *173*, 277 (1978)
45. McCarthy, K. F., Lovenberg, W.: Biochem. Biophys. Res. Commun. *40*, 1053 (1970)
46. Middleton, P., et al.: Eur. J. Biochem. *88*, 135 (1978)
47. Moura, I., et al.: Biochem. Biophys. Res. Commun. *75*, 1037 (1976)
48. Moura, J. J. G., et al.: FEBS Letters *81*, 275 (1977)
49. Moura, J. J. G., et al.: FEBS Letters *89*, 177 (1978)
50. Moura, I., et al.: Biochim. Biophys. Acta *533*, 156 (1978)
51. Moura, I., et al.: FEBS Lett. *107*, 419 (1979)
52. Moura, I., et al.: J. of Biol. Chem. *255*, in press (1980)
53. Mortenson, L. E., Nakos, G.: Bacterial Ferredoxins and/or Iron-Sulfur as Electron Carriers. In: Iron-Sulfur Proteins (ed. Lovenberg, W.), Vol. 1, New York, Academic Press 1973, pp. 37–65
54. Münck, E., et al.: Biochim. Biophys. Acta *400*, 32 (1975)
54a. Münck, E., et al.: Mössbauer Parameters of Putidaredoxin and its Selenium analog, Biochemistry *11*, 855 (1972)
55. Münck, E., Huynh, B. H.: Iron-Sulfur Proteins: Combined Mössbauer and EPR Studies. In: ESR and NMR of Paramagnetic Species in Biological and Related Systems. NATO-Advanced Study Institute (eds. Bertini, I., Drago, R. S., Reidel, D.), Holland 1979, pp. 275–288
56. Ohnishi, T., et al.: J. Biol. Chem. *251*, 2105 (1976)
57. Ohnishi, T., et al.: J. Biol. Chem. *255*, 345 (1980)
58. Orme-Johnson, W. H., Sands, R. H.: Probing Iron-Sulfur Proteins with EPR and ENDOR Spectroscopy. In: Iron-Sulfur Proteins (ed. Lovenberg, W.), Vol. 3, New York, Academic Press 1973, pp. 195–238

212

59. Orme-Johnson, W.H., Davis, L.C.: Current Topics and Problems in the Enzymology of Nitrogenase. In: Iron-Sulfur Proteins (ed. Lovenberg, W.), Vol. 1, New York, Academic Press 1977, pp. 15–60

60. Orme-Johnson, W.H., Holm, R.H.: Identification of Iron Sulfur Clusters in Proteins. In: Methods in Enzymology, Vol. 53 (ed. Fleicher, S., Parker, L.), New York, Academic Press 1978, pp. 631–634

61. Orme-Johnson, W.H., Münck, E.: On the Prosthetic Groups of Nitrogenase. In: Molybdenum and Molybdenum Enzymes (ed. Coughlan, M.), Oxford, Pergamon Press 1980, pp. 429–438

62. Palmer, G., Sands, R.H., Mortenson, L.E.: Biochem. Biophys. Res. Commun. *23*, 357 (1966)

63. Peisach, J., et al.: J. Biol. Chem. *246*, 5877 (1971)

64. Rawlings, J., et al.: J. Biol. Chem. *253*, 1001 (1978)

65. Rendina, A.R., Orme-Johnson, W.H.: Biochem. *15*, 5388 (1978)

66. Rieske, J.S.: Biochim. Biophys. Acta *456*, 195 (1976)

67. Rivoal, J.C., et al.: Biochim. Biophys. Acta *493*, 122 (1977)

68. Ruzicka, F.J., Beinert, H.: J. Biol. Chem. *253*, 2514 (1978)

69. Shah, V.K., Brill, W.J.: Proc. Natl. Acad. Sci. U.S.A. *74*, 3249 (1977)

70. Shethna, Y.I.: Biochim. Biophys. Acta *205*, 58 (1970)

71. Shulman, R.G., et al.: J. Mol. Biol. *124*, 305 (1978)

72. Schulz, C., Debrunner, P.G.: J. Phys. Colloque *37*, Suppl. 12, 153 (1976)

73. Sieker, L.C., et al.: Desulforedoxin: Proposed Configuration and Preliminary X-Ray Diffraction Studies of a two-Iron-Two-Chain Protein, Ciênc. Biol. (1980)

74. Smith, B.E., Lang, G.: Biochem. J. *137*, 169 (1974)

75. Stiefel, E.I.: The Structure and Spectra of Molybdoenzymes Active Sites and their Models. In: Molybdenum and Molybdenum Enzymes (ed. Coughlan, M.), Oxford, Pergamon Press 1980, pp. 43–98

76. Stout, C.D.: Nature *279*, 83 (1979)

77. Stout, C.D., et al.: J. Biol. Chem. *255*, 1797 (1980)

78. Sweeney, W.V., Bearden, A.J., Rabinowitz, J.C.: Biochem. Biophys. Res. Commun. *59*, 188 (1974)

79. Sweeney, W.V., Rabinowitz, J.C.: J. Biol. Chem. *250*, 7842 (1975)

79a. Tsukihara, T.K., et al.: Abstracts of the Sixth International Biophysics Congress, Kyoto, Japan, 1978

80. Watenpaugh, K.D., et al.: Refinement of the Model of a Protein Rubredoxin at 1.5 Å Resolution, Acta Crystallogr. Section B *29*, 943 (1973)

81. Wong, G.B., et al.: J. Am. Chem. Soc. *101*, 3078 (1979)

82. Yang, S.S., et al.: Biochim. Biophys. Acta *590*, 24 (1980)

83. Yates, M.G., et al.: Eur. J. Biochem. *85*, 291 (1978)

84. Yoch, D.C., Arnon, D.I.: J. Biol. Chem. *247*, 4515 (1972)

85. Yoch, D.C., Carithers, R.P.: Microbiological Reviews *43*, 384 (1979)

86. Yoch, D.C., Charithers, R.P.: J. Bacteriol. *136*, 822 (1978)

87. Yoch, D.C., Charithers, R.P., Arnon, D.I.: J. Biol. Chem. *252*, 7453 (1977)

88. Zimmerman, R., et al.: Biochim. Biophys. Acta *537*, 185 (1978)

89. Hardy, R.W., et al.: In: Non Heme Iron Proteins: Role in Energy Conservation (A. San Pietro, ed.), Antioch, Yellow Springs, Ohio 1965, 275

90. Cárdenas, J., Mortenson, L.E., Yoch, D.C.: Purification and Properties of Paramagnetic Protein from *Clostridium pasteurianum* W5, Biochim. Biophys. Acta *434*, 244 (1976)

91. Tagawa, K., Arnon, D.I.: Oxidation-Reduction Potentials and Stoichiometry of Electron Transfer in Ferredoxins, Biochim. Biophys. Acta *153*, 602 (1968)

92. Shethna, Y.I., Stombaugh, N.A., Burris, R.H.: Ferredoxin from *Bacilus polymyxa*, Biochem. Biophys. Res. Commun. *42*, 1108 (1971)

Author-Index Volumes 1—43

215

THEORETICA CHIMICA ACTA

an International Journal of Theoretical Chemistry

ISSN 0040-5744   TitleNo. 214

Edenda curat: Hermann Hartmann, Mainz

Adiuvantibus: C. J. Ballhausen, København; R. D. Brown, Clayton; K. Fukui, Kyoto; R. Gleiter, Heidelberg; E. A. Halevi, Haifa; G. G. Hall, Nottingham; E. Heilbronner, Basel; J. Jortner, Tel-Aviv; M. Kotani, Tokyo; J. Koutecký, Berlin; A. Neckel, Wien; E. E. Nikitin, Moskwa; R. G. Pearson, Santa Barbara; B. Pullmann, Paris; B. Rånby, Stockholm; K. Ruedenberg, Ames; C. Sandorfy, Montreal; M. Simonetta, Milano; O. Sinanoğlu, New Haven; R. Zahradník, Praha

Today, theory and experiment are inseparably bound. Every chemical experiment is preceded by reflection and careful consideration, and the results are interpreted according to chemical theories and perceptions.

The editors of **Theoretica Chimica Acta** therefore wish to emphasize the wide-ranging program reflected in the policy of their journal:

"**Theoretica Chimica Acta** accepts manuscripts in which the relationships between individual chemical and physical phenomena are investigated. In addition, experimental research that presents new theoretical viewpoints is desired."

Theoretica Chimica Acta offers experimental chemists increased space for the publication of discussion of the goals of their work, the significance of their findings, and the concepts on which their experimental work is based. Such discussions contribute significantly to mutual understanding between theoreticians and experimentalists and stimulate both new reflections and further experiments.

Springer International

Subscription Information and/or sample copies upon request. Please send your order or request to your bookseller or directly to:
Springer-Verlag, Journal Promotion Department, P. O. Box 105280, D-6900 Heidelberg, FRG

Lecture Notes in Chemistry

Edited by G. Berthier, M. J. S. Dewar, H. Fischer,
K. Fukui, H. Hartmann, H. H. Jaffé, J. Jortner,
W. Kutzelnigg, K. Ruedenberg, E. Scrocco, W. Zeil

Springer-Verlag
Berlin
Heidelberg
New York